Shunji Osaki

Applied Stochastic System Modeling

With 55 Figures

Springer-Verlag

Berlin Heidelberg New York
London Paris Tokyo
Hong Kong Barcelona
Budapest

Professor Dr. Shunji Osaki
Department of Industrial and
Systems Engineering
Faculty of Engineering
Hiroshima University
Higashi-Hiroshima 724
Japan

ISBN-13:978-3-642-84683-0 e-ISBN-13:978-3-642-84681-6
DOI: 10.1007/978-3-642-84681-6

Softcover reprint of the hardcover 1st edition 1992

2142/7130-543210 – Printed on acid-free paper

Preface

This book was written for an introductory one-semester or two-quarter course in stochastic processes and their applications. The reader is assumed to have a basic knowledge of analysis and linear algebra at an undergraduate level. Stochastic models are applied in many fields such as engineering systems, physics, biology, operations research, business, economics, psychology, and linguistics. Stochastic modeling is one of the promising kinds of modeling in applied probability theory.

This book is intended to introduce basic stochastic processes: Poisson processes, renewal processes, discrete-time Markov chains, continuous-time Markov chains, and Markov-renewal processes. These basic processes are introduced from the viewpoint of elementary mathematics without going into rigorous treatments. This book also introduces applied stochastic system modeling such as reliability and queueing modeling.

Chapters 1 and 2 deal with probability theory, which is basic and prerequisite to the following chapters. Many important concepts of probabilities, random variables, and probability distributions are introduced.

Chapter 3 develops the Poisson process, which is one of the basic and important stochastic processes. Chapter 4 presents the renewal process. Renewal-theoretic arguments are then used to analyze applied stochastic models. Chapter 5 develops discrete-time Markov chains. Following Chapter 5, Chapter 6 deals with continuous-time Markov chains. Continuous-time Markov chains have important applications to queueing models as seen in Chapter 9. A one-semester course or two-quarter course consists of a brief review of Chapters 1 and 2, followed in order by Chapters 3 through 6.

Chapter 7 develops the Markov renewal processes. Chapter 8 presents reliability models. Finally, Chapter 9 develops queueing models. Chapters 7, 8, and 9 are devoted to applied stochastic system modeling. For interested readers, Chapters 7, 8, and 9 discuss thought-provoking areas of applied stochastic system modeling.

Appendix A is devoted to the Laplace-Stieltjes transforms. Appendix B is devoted to answers to selected problems. Appendix C contains a bibliography.

I would like to thank Professor Naoto Kaio, Hiroshima Shudo University, who has read an earlier draft of the manuscript and corrected misunderstandings and mistakes. Special thanks go to several students from my laboratory at Hiroshima University, who provided rapid and excellent word processing and figure drawing. This book was prepared with TEX[1], and numerical computations in this book were carried out by *Mathematica*[2]. I would finally like to thank Dr. Werner A. Müller, Economics Editor, Springer-Verlag, for his kindness and encouragement.

Shunji Osaki

Higashi-Hiroshima, Japan

[1]TEX is a trademark of the American Mathematical Society.
[2]*Mathematica* is a trademark of Wolfram Research Inc.

Contents

Chapter 1

Probability Theory

1.1 Introduction

We cannot predict in advance the outcome of tossing a coin or a die, drawing a card, and so on. Such an experiment is called a random trial since nobody knows the outcome in advance and it seems to be "random". However, we can know all the possible outcomes of a random trial in advance. For instance, if we consider tossing a coin, the possible outcomes are "H(heads)" and "T(tails)". Knowing all the possible outcomes in advance, we should assign each outcome with a number in advance; this is its probability.

Probability theory is a field of mathematics treating random phenomena quantitatively. We briefly sketch probability theory in Chapters 1 and 2 as a basis for subsequent discussions. We do not intend to treat probability theory from the viewpoint of measure theory, but rather from that of elementary mathematics at the expense of rigorous treatments. The reader should consult other textbooks on probability theory for a rigorous presentation.

1.2 Sample Spaces and Events

We shall now consider an experiment of tossing a coin or a die, drawing a card, and so on, in which the outcome of the experiment is not predictable in advance, but all the possible outcomes are predictable in advance, and introduce elementary terms in probability theory. Consider a *random trial* such as coin tossing, die tossing, and card drawing. The outcome of the random trial is called a *sample point* (or simply *point*). The set of all the possible outcomes of the trial is called a *sample space*. An *event* is a subset of the sample space. We usually denote a sample point by ω (Greek small omega), an event by any capital letter (say, A, B, $\cdots$), and a sample space by Ω (Greek capital omega).

Example 1.2.1 We show some examples of the sample spaces.

(1) If we consider a random trial of tossing (flipping) a coin, then

$$\Omega = \{H, T\},$$

where H and T mean that the outcomes are "heads" and "tails", respectively.

(2) If we consider a random trial of tossing a die, then

$$\Omega = \left\{\boxed{1}, \boxed{2}, \boxed{3}, \boxed{4}, \boxed{5}, \boxed{6}\right\},$$

where $\boxed{i}$ means that the outcome is i on the die, $i = 1, 2, 3, 4, 5, 6$.

(3) A deck of (bridge) cards consists of 52 cards arranged in four suits of thirteen cards each. There are thirteen face values (A(ce), 2, 3, 4, 5, 6, 7, 8, 9, 10, J(ack), Q(ueen), and K(ing)) in each suit. The four suits are $\spadesuit$ (spades), $\clubsuit$ (clubs), $\heartsuit$ (hearts), and $\diamondsuit$ (diamonds), where $\spadesuit$ and $\clubsuit$ are black, and $\heartsuit$ and $\diamondsuit$ are red. Cards of the same face value are considered the same. If we consider a random trial of drawing a card from an ordinary deck of 52 cards, then

$$\Omega = \{T_i\,;\ T = \spadesuit, \clubsuit, \heartsuit, \diamondsuit\,;\ i = A, 2, \cdots, 10, J, Q, K\},$$

where T_i means that the outcome is the (face value) i of (suit) T. For instance, $\spadesuit_A$ means that the outcome is the Ace of spades, and $\heartsuit_7$ means that the outcome is the 7 of hearts.

(4) If we consider a random trial of measuring the height of a person, then the sample space [1] is

$$\Omega = \{x;\ x \in [0, \infty)\}.$$

The above examples (1), (2), and (3) are *discrete* sample spaces, but the last example (4) is a *continuous* sample space.

Example 1.2.2 We present other complicated examples of tossing a coin many times.

(1) If we consider a random trial of tossing a coin twice, then

$$\Omega = \{(H, H), (H, T), (T, H), (T, T)\}.$$

In particular, if A is an event where heads appears first, and B is an event where heads appears second, then the two events A and B are

$$A = \{(H, H), (H, T)\}, \text{ and } B = \{(H, H), (T, H)\}.$$

[1] The set $[0, \infty)$ is defined to consist of all points x such that $0 \le x < \infty$. In general, similar notations are used (a, b), $[a, b)$, $(a, b]$, and $[a, b]$ meaning all points x such that $a < x < b$, $a \le x < b$, $a < x \le b$, and $a \le x \le b$, respectively, where a might be $-\infty$, and b might be $+\infty$.

(2) If we consider a random trial of tossing a coin three times, then

$$\Omega = \{(H,\ H,\ H),\ (H,\ H,\ T),\ (H,\ T,\ H),\ (T,\ H,\ H),$$

$$(H,\ T,\ T),\ (T,\ H,\ T),\ (T,\ T,\ H),\ (T,\ T,\ T)\},$$

where the number of all possible outcomes (sample points) is $2 \cdot 2 \cdot 2 = 2^3 = 8$.

(3) If we consider a random trial of tossing a coin n times, then the number of all possible outcomes (sample points) is $\overbrace{2 \cdot 2 \cdot 2 \cdots 2}^{n} = 2^n$ (*multiplication principle*).

In Example 1.2.2 (1), we have shown examples of events A and B in the sample space Ω. We have described the operations of the events. These events can be performed by adopting the operations of the sets. It is noted that a sample space Ω itself is an event and the *null* (*empty* or *impossible*) *set* ϕ is an event where the null set ϕ contains no sample points.

Definition 1.2.1 For any events A and B in a sample space Ω, we define

(i) Union $A \cup B = \{\omega : \omega \in A \text{ or } \omega \in B\}$.

(ii) Intersection $AB = A \cap B = \{\omega : \omega \in A \text{ and } \omega \in B\}$.

(iii) Complement $A^c = \{\omega : \omega \notin A\}$.

(iv) Exclusion $AB = \phi$.

(v) Inclusion $A \subset B$ (not excluding $A = B$), i.e., any sample point $\omega \in A$ belongs to B (the event A is a subset of the event B).

(vi) Difference $A - B = AB^c = \{\omega : \omega \in A \text{ and } \omega \notin B\}$.

We prefer to indicate the intersection of two events A and B by AB rather than by $A \cap B$ throughout this book. In particular, if (iv) holds, the events A and B are said to be *mutually exclusive*. A sample space Ω is called the *total event* and ϕ the *null event*. Figure 1.2.1 shows the operations of the events by using the so-called *Venn diagrams.*

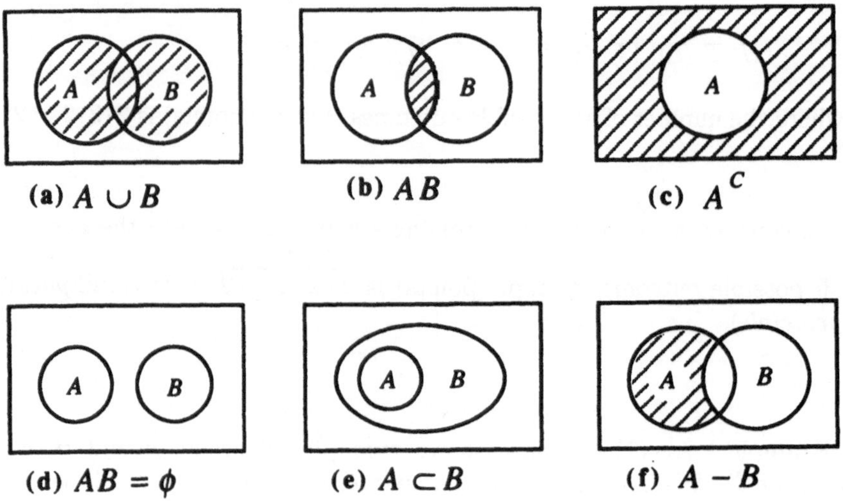

Fig. 1.2.1 Venn diagrams of the operations of the events.

Example 1.2.3 Consider a random trial of tossing a die as in Example 1.2.1 (2). A_1, A_2, B_1, and B_2 denote the events of odd numbers, even numbers, 1, 2, or 3, and 4, 5, or 6, respectively. Then

$$A_1 = \{\boxed{1}, \boxed{3}, \boxed{5}\}, \qquad A_2 = \{\boxed{2}, \boxed{4}, \boxed{6}\},$$

$$B_1 = \{\boxed{1}, \boxed{2}, \boxed{3}\}, \qquad B_2 = \{\boxed{4}, \boxed{5}, \boxed{6}\}.$$

The operations of the events above are

$$A_1 \cup A_2 = \Omega, \qquad A_1 A_2 = \phi,$$

$$B_1 \cup B_2 = \Omega, \qquad B_1 B_2 = \phi,$$

$$A_1 \cup B_1 = \{\boxed{1}, \boxed{2}, \boxed{3}, \boxed{5}\},$$

$$A_2 \cup B_1 = \{\boxed{1}, \boxed{2}, \boxed{3}, \boxed{4}, \boxed{6}\},$$

$$A_1 \cup B_2 = \{\boxed{1}, \boxed{3}, \boxed{4}, \boxed{5}, \boxed{6}\},$$

$$A_2 \cup B_2 = \{\boxed{2}, \boxed{4}, \boxed{5}, \boxed{6}\},$$

$$A_1 B_1 = \{\boxed{1}, \boxed{3}\}, \qquad A_2 B_1 = \{\boxed{2}\},$$

$$A_1 B_2 = \{\boxed{5}\}, \qquad A_2 B_2 = \{\boxed{4}, \boxed{6}\}.$$

Note that the events A_1 and A_2, and B_1 and B_2 are mutually exclusive. Note also that A_1 and A_2, and B_1 and B_2 are mutually exclusive and exhaustive since $A_1 \cup A_2 = B_1 \cup B_2 = \Omega$ (the total event).

Example 1.2.4 Let Ω denote a sample space of students at Hiroshima University. Let A_1, A_2, A_3, and A_4 denote the events of freshmen, sophomores, juniors, and seniors, respectively. Further, let M and F denote the events of male and female students, respectively.

(1) The event of male students among freshmen or sophomores is $(A_1 \cup A_2)M$.

(2) The event of male students among seniors or female students among freshmen is $A_4M \cup A_1F$.

(3) It is clear that

$$A_1 \cup A_2 \cup A_3 \cup A_4 = M \cup F = \Omega,$$

$$A_iA_j = \phi \qquad (i \neq j), \qquad MF = \phi,$$

i.e., the events A_1, A_2, A_3, and A_4, and M and F are mutually exclusive and exhaustive.

Example 1.2.5 Referring to Fig. 1.2.1 (b), we show that

$$AB \subset B, \qquad AB \subset A,$$

$$A \cup B = A \cup (B - AB), \qquad A(B - AB) = \phi,$$

$$A \cup B = B \cup (A - AB), \qquad B(A - AB) = \phi,$$

i.e., AB is a subset of A or B, and A and $B - AB$, and B and $A - AB$ are mutually exclusive.

1.3 Probabilities

We have discussed events in a sample space Ω and the operations of these events in the preceding section. In this section, we introduce the probability of the event A.

Definition 1.3.1 For each event A in a sample space Ω, a real number $P\{A\}$ is defined and satisfied by the following three axioms:

(i) $0 \leq P\{A\} \leq 1.$ (1.3.1)

(ii) $P\{\Omega\} = 1.$ (1.3.2)

(iii) For any sequence of events $A_1, A_2, \cdots$ that are mutually exclusive,

$$P\left\{\bigcup_{n=1}^{\infty} A_n\right\} = \sum_{n=1}^{\infty} P\{A_n\}, \tag{1.3.3}$$

where $P\{A\}$ is called the *probability* of the event A.

Example 1.3.1 We show how to calculate the probabilities for Example 1.2.1.

(1) Consider Example 1.2.1 (1). If the coin is "fair" (i.e., symmetrical), then

$$P\{H\} = P\{T\} = \frac{1}{2}.$$

(2) For Example 1.2.1 (2), if the die is fair, then

$$P\{\boxed{i}\} = \frac{1}{6} \qquad (i = 1, 2, 3, 4, 5, 6).$$

The first example is often called the *fair* (or *symmetrical*) coin. That is, if all the possible outcomes are equally likely to appear, then we call such trials *equally likely* trials. In general, we assume such fair coin and die unless otherwise specified.

(3) In terms of Example 1.2.1 (3), if we randomly draw a card out of a deck of 52 (i.e., the equally likely trial), then

$$P\{T_i\} = \frac{1}{52}.$$

As shown in the example above, we can assign each event a probability that satisfies the three axioms in Definition 1.3.1. For instance, we can examine the probability that heads appear r times by trying n tosses ($r \leq n$) for a "fair" coin in practice. Then we can calculate the ratio $f = r/n$, the "relative frequency." However, we cannot guarantee $f = r/n$ tends to $1/2$ as n tends to infinity. We can only assign each probability axiomatically. The problem, of estimating and testing probability and other quantities, remains a subject of "statistics."

Example 1.3.2 Calculate the probabilities for Example 1.2.2.

(1) Consider Example 1.2.2 (1). Then

$$P\{(i, j)\} = \frac{1}{4} \qquad (i, j = H, T).$$

(2) For Example 1.2.2 (2), we have

$$P\{(i, j, k)\} = \frac{1}{8} \qquad (i, j, k = H, T).$$

Applying Definition 1.3.1 and the results in the preceding section, we present the following theorem (see Fig. 1.2.1 for proofs).

Theorem 1.3.1 For any events A and B, the following hold:

(i) $P\{A^c\} = 1 - P\{A\}$, in particular, $P\{\phi\} = 0$. (1.3.4)

(ii) $P\{A - B\} = P\{A\} - P\{AB\}$. (1.3.5)

(iii) If $A \subset B$, then $P\{A\} \le P\{B\}$. (1.3.6)

(iv) $P\{A \cup B\} = P\{A\} + P\{B\} - P\{AB\}$. (1.3.7)

Proof

(i) Since $A \cup A^c = \Omega$, and $AA^c = \phi$ (i.e., the events A and A^c are mutually exclusive and exhaustive), we have

$$1 = P\{\Omega\} = P\{A \cup A^c\} = P\{A\} + P\{A^c\}. \tag{1.3.8}$$

In particular, if $A = \Omega$, then $A^c = \phi$.

(ii) From Definition 1.2.1 (vi), we show that the events $A - B$ and AB are mutually exclusive and $(A - B) \cup AB = A$. Then

$$P\{A\} = P\{(A - B) \cup AB\} = P\{A - B\} + P\{AB\}. \tag{1.3.9}$$

(iii) If $A \subset B$, then B can be decomposed into the mutually exclusive events A and $B - A$. Then

$$\begin{aligned} P\{B\} &= P\{A \cup (B - A)\} = P\{A\} + P\{B - A\} \\ &\ge P\{A\}, \end{aligned} \tag{1.3.10}$$

since $P\{B - A\} \ge 0$.

(iv) The event $A \cup B$ can be decomposed into the mutually exclusive events $A - B$ and B. Then

$$\begin{aligned} P\{A \cup B\} &= P\{(A - B) \cup B\} \\ &= P\{A - B\} + P\{B\} \\ &= P\{A\} - P\{AB\} + P\{B\}, \end{aligned} \tag{1.3.11}$$

since $P\{A - B\} = P\{A\} - P\{AB\}$ from Eq.(1.3.5).

Applying Theorem 1.3.1 (iv) twice, we have the following corollary:

Corollary 1.3.1 For any events A, B, and C,

$$P\{A \cup B \cup C\} = P\{A\} + P\{B\} + P\{C\}$$
$$-P\{AB\} - P\{BC\} - P\{AC\} + P\{ABC\}. \qquad (1.3.12)$$

The proof is left for exercise. Consult Fig. 1.3.1 for understanding Eq.(1.3.12).

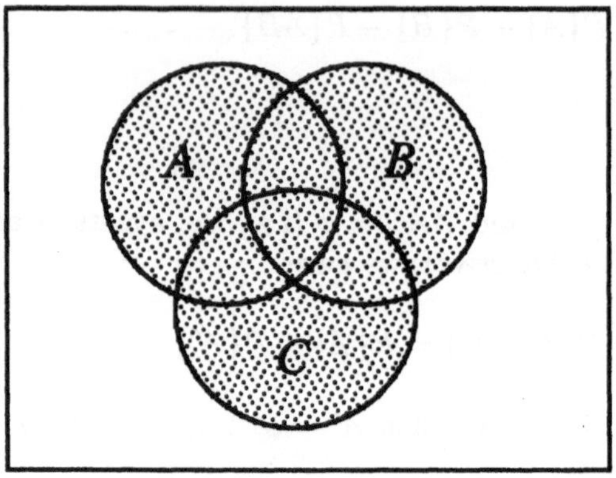

Fig. 1.3.1 The event $A \cup B \cup C$.

Example 1.3.3 The students can pass courses A and B with the following probabilities: A : 50%, B : 40%, AB : 10%.

(1) The probability of passing at least one course is

$$P\{A \cup B\} = P\{A\} + P\{B\} - P\{AB\} = 0.5 + 0.4 - 0.1 = 0.8.$$

(2) The respective probabilities of passing only course A or B are

$$P\{A - AB\} = P\{A\} - P\{AB\} = 0.5 - 0.1 = 0.4,$$

$$P\{B - AB\} = P\{B\} - P\{AB\} = 0.4 - 0.1 = 0.3.$$

We now introduce conditional probability.

Definition 1.3.2 For any events A and B in a sample space Ω, the *conditional probability* of the event A given the event B is defined by

$$P\{A \mid B\} = \frac{P\{AB\}}{P\{B\}}, \qquad (1.3.13)$$

if $P\{B\} > 0$. Equation (1.3.13) can be written by

$$P\{AB\} = P\{A \mid B\}\, P\{B\}. \tag{1.3.14}$$

If

$$P\{AB\} = P\{A\}\, P\{B\} \tag{1.3.15}$$

or equivalently

$$P\{A \mid B\} = P\{A\}, \tag{1.3.16}$$

then the events A and B are said to be *mutually independent* or simply *independent.*

Example 1.3.4 (*Continuation* of Example 1.3.3)

(1) The conditional probability that the students pass the two courses A and B given that they pass the course A is

$$P\{AB \mid A\} = \frac{P\{AB\}}{P\{A\}} = \frac{1}{5}.$$

(2) The similar conditional probability for passing course B is

$$P\{AB \mid B\} = \frac{P\{AB\}}{P\{B\}} = \frac{1}{4}.$$

(3) The conditional probability that the students pass both courses A and B given that they pass at least one course is

$$P\{AB \mid A \cup B\} = \frac{P\{AB\}}{P\{A \cup B\}} = \frac{1}{8}.$$

In Definition 1.3.2, we show that

$$P\{AB\} = P\{A \mid B\}\, P\{B\} \tag{1.3.17}$$

for any two events A and B, and $P\{B\} > 0$. Just as in Eq.(1.3.17), we can easily generalize to n events in the following corollary:

Corollary 1.3.2 For any events $A_1,\ A_2,\ \cdots,\ A_n$,

$$P\{A_1A_2\cdots A_n\} = P\{A_1\}P\{A_2 \mid A_1\}P\{A_3 \mid A_1A_2\} \cdots P\{A_n \mid A_1A_2\cdots A_{n-1}\}. \tag{1.3.18}$$

We can easily generalize the independent events for any n events.

Definition 1.3.3 For any n events $A_1, A_2, \cdots, A_n$, if

$$P\{A_{i_1} A_{i_2} \cdots A_{i_k}\} = P\{A_{i_1}\} P\{A_{i_2}\} \cdots P\{A_{i_k}\} \tag{1.3.19}$$

for $1 \leq i_1 < i_2 < \cdots < i_k \leq n$, $2 \leq k \leq n$, then the events $A_1, A_2, \cdots, A_n$ are *mutually independent* or simply *independent*.

Example 1.3.5 To verify that the events A, B, and C are mutually independent, we should present the following identities:

$$P\{AB\} = P\{A\} P\{B\}, \qquad P\{BC\} = P\{B\} P\{C\},$$

$$P\{AC\} = P\{A\} P\{C\}, \qquad P\{ABC\} = P\{A\} P\{B\} P\{C\}.$$

Example 1.3.6 We consider a random trial of tossing a die twice. Let A, B, and C denote the event that the an odd-numbered face appears first, an odd-numbered face appears second, and the sum of the two faces is odd, respectively. Since $P\{A\} = P\{B\} = P\{C\} = \frac{1}{2}$, we have

$$P\{AB\} = P\{A\} P\{B\} = \frac{1}{4}, \qquad P\{BC\} = P\{B\} P\{C\} = \frac{1}{4},$$

$$P\{AC\} = P\{A\} P\{C\} = \frac{1}{4}, \qquad P\{ABC\} = 0, \quad P\{A\} P\{B\} P\{C\} = \frac{1}{4^3},$$

where $ABC = \phi$ (i.e., it is impossible that A, B, and C occur simultaneously). That is, any two of the events are independent, but the event that all the events occur simultaneously is impossible.

We often cited the example of tossing a coin or a die twice, three times, and so on. This concept of repetition is defined as follows:

Definition 1.3.4 (*Independent* or *repeated trials*) Let $A_1, A_2, \cdots, A_n$ be n random trials for a sample space Ω. The random trials are called *stochastically independent* or *statistically independent*, or simply *independent* if for every $A_i \subset \Omega$ $(i = 1, 2, \cdots, n)$,

$$P\{A_1 A_2 \cdots A_n\} = P\{A_1\} P\{A_2\} \cdots P\{A_n\}. \tag{1.3.20}$$

Example 1.3.7 Successive tossing of a coin or a die is a typical example of stochastic independence. Let us consider Example 1.2.2 (3). The probability that all n trials of tossing a coin n times appear H simultaneously is

$$P\{(\overbrace{HH \cdots H}^{n})\} = \frac{1}{2^n},$$

since they are independent trials.

We have cited a few examples where events are mutually exclusive and exhaustive. In this context, we introduce the following concepts:

Definition 1.3.5 The events $A_1,\ A_2,\ \cdots,\ A_n$ form a *partition* of a sample space Ω, if the events A_i are mutually exclusive and their union is Ω. That is, if

$$A_iA_j = \phi \tag{1.3.21}$$

for any $i \neq j$, and

$$A_1 \cup A_2 \cup \cdots \cup A_n = \Omega, \tag{1.3.22}$$

then the events $A_1, A_2, \cdots, A_n$ form a partition.

The following theorem is well-known as the total probability formula:

Theorem 1.3.2 (*Total Probability Formula*) Let $A,\ B_1,\ B_2,\ \cdots,\ B_n$ denote the events such that $B_iB_j = \phi$ for any $i \neq j$ and

$$A \subset (B_1 \cup B_2 \cup \cdots \cup B_n). \tag{1.3.23}$$

Then

$$P\{A\} = \sum_{i=1}^{n} P\{A \mid B_i\}\, P\{B_i\}, \tag{1.3.24}$$

where $P\{B_i\} > 0$ $(i = 1,\ 2,\ \cdots,\ n)$.

Proof Noting that

$$P\{A \mid B_i\}\, P\{B_i\} = P\{AB_i\} \tag{1.3.25}$$

and

$$(AB_i)(AB_j) = A(B_iB_j) = \phi, \tag{1.3.26}$$

for any $i \neq j$, we have

$$\begin{aligned} &\sum_{i=1}^{n} P\{A \mid B_i\}\, P\{B_i\} \\ &= \sum_{i=1}^{n} P\{AB_i\} \\ &= P\{AB_1 \cup AB_2 \cup \cdots \cup AB_n\} \\ &= P\{A(B_1 \cup B_2 \cup \cdots \cup B_n)\} \\ &= P\{A\}. \end{aligned} \tag{1.3.27}$$

Note that the above theorem is valid if the events $B_1,\ B_2,\ \cdots,\ B_n$ form a partition, since it is always true that

$$A \subset (B_1 \cup B_2 \cup \cdots \cup B_n) = \Omega. \tag{1.3.28}$$

Using the so-called total probability formula, we are now ready to show the following:

Theorem 1.3.3 (*Bayes' Theorem*) Let $A,\ B_1,\ B_2,\ \cdots,\ B_n$ denote the events such that $B_iB_j = \phi$ for any $i \neq j$ and

$$A \subset (B_1 \cup B_2 \cup \cdots \cup B_n). \tag{1.3.29}$$

Then

$$P\{B_i \mid A\} = \frac{P\{A \mid B_i\}\,P\{B_i\}}{\sum_{i=1}^{n} P\{A \mid B_i\}\,P\{B_i\}}, \tag{1.3.30}$$

where $P\{A\} > 0$ and $P\{B_i\} > 0$ for $i = 1,\ 2,\ \cdots,\ n$.

Proof Using the conditional probability in Definition 1.3.2, we have

$$P\{B_i \mid A\} = \frac{P\{B_iA\}}{P\{A\}} = \frac{P\{A \mid B_i\}\,P\{B_i\}}{P\{A\}}, \tag{1.3.31}$$

and we adopt the total probability formula for the denominator of the right-hand side of the above equation.

Example 1.3.8 A factory produces the same items by using three machines, say B_1, B_2, and B_3, whose production capacities are 60%, 30%, and 10%, respectively. The percentages of defective items each machine produces are 6%, 3%, and 5%, respectively.

(1) Calculate the probability that a randomly selected item is defective. Let A be the event that a randomly selected item is defective. Then, applying the total probability formula in Eq.(1.3.24), we have

$$\begin{aligned} P\{A\} &= \sum_{i=1}^{3} P\{A \mid B_i\}\,P\{B_i\} \\ &= 0.06 \times 0.60 + 0.03 \times 0.30 + 0.05 \times 0.10 \\ &= 0.05. \end{aligned}$$

(ii) Calculate the conditional probabilities that randomly selected defective items were produced by machines B_1, B_2, and B_3, respectively. Applying the Bayes' formula in Eq.(1.3.30), we have

$$P\{B_1 \mid A\} = \frac{P\{A \mid B_1\}\,P\{B_1\}}{P\{A\}} = \frac{36}{50} = 72\%,$$

$$P\{B_2 \mid A\} = \frac{P\{A \mid B_2\} P\{B_2\}}{P\{A\}} = \frac{9}{50} = 18\%,$$

$$P\{B_3 \mid A\} = \frac{P\{A \mid B_3\} P\{B_3\}}{P\{A\}} = \frac{5}{50} = 10\%.$$

1.4 Combinatorial Analysis

In the preceding section we have introduced probability and showed several examples of calculating probabilities. To calculate probability, we should consider enumerating the number of possible outcomes of a random trial. In this section we focus on the counting technique, i.e., so-called *combinatorial analysis*.

Let a be a real number and r be a non-negative integer. Then we define

$$(a)_r = a(a-1)(a-2)\cdots(a-r+1), \tag{1.4.1}$$

where $(a)_r$ is the product of r numbers from a to $a - r + 1$ downwards. In particular, we postulate

$$(a)_0 = 1, \tag{1.4.2}$$

for $r = 0$.

If a is assumed to be a non-negative integer and $a = r$, we have

$$(r)_r = r(r-1)(r-2)\cdots 2 \cdot 1 = r!, \tag{1.4.3}$$

where

$$(0)_0 = 0! = 1. \tag{1.4.4}$$

The special symbol $r!$ (read "r factorial") plays a central role for combinatorial analysis.

Again we assume in general that a is a real number and r is a non-negative integer. We define the *binomial coefficient*

$$\binom{a}{r} = {}_aC_r = \frac{(a)_r}{r!} = \frac{a(a-1)\cdots(a-r+1)}{r!} \qquad (r > 0), \tag{1.4.5}$$

and

$$\binom{a}{0} = 1 \qquad (r = 0). \tag{1.4.6}$$

In particular, if $a = n$ and r are integers such that $n \geq r \geq 0$, then

$$(n)_r = n(n-1)\cdots(n-r+1) = \frac{n!}{(n-r)!}, \tag{1.4.7}$$

$$\binom{n}{r} = \frac{n(n-1)\cdots(n-r+1)}{r!} = \frac{n!}{r!(n-r)!}. \tag{1.4.8}$$

Additionally, if r is a negative integer, then we postulate

$$\binom{a}{r} = 0. \tag{1.4.9}$$

The following theorem is basic for determining the number of different ways to perform an operation in combinatorial analysis.

Theorem 1.4.1 (*Multiplication Principle*) If an operation A_1 can be performed in n_1 different ways, another operation A_2 in n_2 different ways, etc., the kth operation A_k in n_k different ways, then the k operations can be performed one after another in $n_1 \cdot n_2 \cdots n_k$ ways.

Proof The proof is performed by considering the Cartesian product

$$A_1 \times A_2 \times \cdots \times A_k = \{(x_1, x_2, \cdots, x_k) : x_i \in A_i, i = 1, 2, \cdots, k\}$$

and the size of the above product is $n_1 \cdot n_2 \cdots n_k$.

We use the word "sampling" for tossing a coin or a die, drawing a card, selecting a number, a key, or a door, and so on. We have the following two types of sampling.

Definition 1.4.1 Consider the set which is composed of n different objects $a_1, a_2, \cdots, a_n$. Any ordered arrangement $(a_{j_1}, a_{j_2}, \cdots, a_{j_n})$ of n symbols is called an *ordered sample of size n* taken from the set. *Sampling with replacement* is defined by the fact that each selection is made from the set of all objects (i.e., repetitions are allowed). *Sampling without replacement* is defined by the fact that an object once chosen is removed from the set (i.e., repetitions are not allowed).

In the definition above, any arrangement of the ordered sample of size r without replacement is called an *r-permutation* or simply a *permutation.*

The following theorem is basic for determining the number of different ways to perform an operation in general.

Theorem 1.4.2 Let n and r be integers such that $n \geq r \geq 0$.

(i) The number of different ordered samples of size r taken from the set of n objects with replacement is n^r.

(ii) The number of different ordered samples of size r taken from the set of n objects without replacement (i.e., the r-permutation) is $(n)_r$.

(iii) The number of different *combinations* (ignoring ordering) of size r taken from the set of n objects without replacement is $\binom{n}{r}$.

(iv) The number of different *combinations* (ignoring ordering) of size r taken from the set of n objects with replacement is

$$\binom{n+r-1}{n-1} = \binom{n+r-1}{r}. \tag{1.4.10}$$

Proof

(i) The proof is performed by applying the Multiplication Principle from Theorem 1.4.1.

(ii) By applying the Multiplication Principle and noting that the set of the objects decreases one by one, we have $(n)_r$ ways.

(iii) The number of different ways of r-permutations is $(n)_r$ (from (ii)). However, if we ignore their order, we have $(r)_r = r!$ permutations which are the same in combination. Thus, we have $\binom{n}{r} = (n)_r/r!$ combinations of size r taken from the set of n objects.

(iv) Let n objects denote $1, 2, \cdots, n$. Allowing the repetitions, we have r-tuples, e.g.,

$$\overbrace{1,\ 1,\ 3,\ 3,\ 4,\ 5,\ 5,\ 6,\ \cdots,\ n}^{r},$$

where these numbers are arranged in ascending order. If the adjacent numbers are different in the r-tuples, insert $*$ or $*$'s as the difference of the two adjacent numbers, e.g.,

$$1,\ 1,\overbrace{*,\ *,}^{2}\ 3,\ 3,\ *,\ 4,\ *,\ 5,\ 5,\ *,\ 6,\ \cdots,\ n.$$

Then, we can insert $*$'s $(n-1)$ times in the $(n+r-1)$-tuples. Once, if we can allocate $*$'s in the $(n+r-1)$-tuples, we can generate the r-tuples which correspond to all the different ways of the r combinations taken from the set of n objects with replacement. It is clear that the number of allocating $*$'s in the $(n+r-1)$-tuples is equivalent to considering $\binom{n+r-1}{n-1}$ ways of combinations since we ignore the ordering of $(n-1)*$'s (see Example 1.4.1 (4)).

Example 1.4.1 Consider the set of 3 objects (say a, b, c). Determine the number of different ways for $r = 3$ and enumerate them in terms of the above theorem.

(1) $3^3 = 27$ ways.

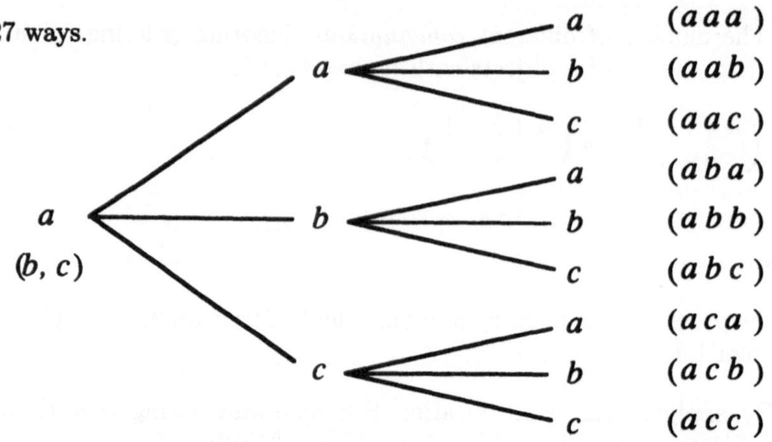

Not shown are the remaining orderings where b or c appears first (change the first element in the above parentheses to b or c).

(2)

$(3)_3 = 3! = 6$ ways, and

(abc), (acb), (bac), (bca), (cab), (cba).

(3)

$$\binom{3}{3} = \frac{3!}{3!} = 1 \text{ way, and } (abc).$$

If we consider the number of different combinations of 2 elements taken from the set of 3 objects without replacement,

$$\binom{3}{2} = \frac{3 \cdot 2}{1 \cdot 2} = 3 \text{ ways, and } (ab),\ (bc),\ (ac).$$

(4)

$$\binom{3+3-1}{3-1} = \binom{5}{2} = \frac{5 \cdot 4}{1 \cdot 2} = 10 \text{ ways, and}$$

(aaa), (bbb), (ccc), (aab), (abb),

(bbc), (bcc), (aac), (acc), (abc).

$(45) \to (aaa)$, $(15) \to (bbb)$, $(12) \to (ccc)$,

$(35) \to (aab)$, $(25) \to (abb)$, $(14) \to (bbc)$,

$(13) \to (bcc)$, $(34) \to (aac)$, $(23) \to (acc)$,

$(24) \to (abc)$.

The following are the elementary identities in binomial coefficients.

Theorem 1.4.3 Let n and r be integers such that $n \geq r \geq 0$.

(i)

$$\binom{n}{r} = \binom{n}{n-r}. \tag{1.4.11}$$

(ii)

$$\binom{n+1}{r} = \binom{n}{r-1} + \binom{n}{r}. \tag{1.4.12}$$

(iii)

$$\binom{n+r-1}{r-1} = (-1)^{r-1}\binom{-n-1}{r-1}. \tag{1.4.13}$$

(iv)

$$(a+b)^n = \sum_{r=0}^{n} \binom{n}{r} a^r b^{n-r}. \tag{1.4.14}$$

Example 1.4.2 (*Birthday Problem*) There are r students in a class. If we assume that each student's birthday is equally likely and we ignore leap years, what is the probability that all the r students have distinct birthdays? There are 365^r ways in which r students have their birthdays. We have $(365)_r$ ways of all r distinct birthdays, where $r \leq 365$. That is, the probability that all r students have distinct birthdays is

$$P = \frac{(365)_r}{365^r} = \left(1 - \frac{1}{365}\right)\left(1 - \frac{2}{365}\right)\cdots\left(1 - \frac{r-1}{365}\right).$$

Table 1.4.1 shows the computational results for $r = 11, \cdots, 30$. From this table we see that, if there are 23 students in a classroom, the probability that at least two students have the same birthday exceeds 1/2. It is tedious to calculate the probability above. We show a rough approximation for small r. Expanding the right-hand side of P, we approximate the two terms of P:

$$P \approx 1 - \frac{1 + 2 + \cdots + (r-1)}{365} = 1 - \frac{r(r-1)}{730}.$$

Table 1.4.1 also shows the above approximations which are reasonably useful in practice for small r.

Theorem 1.4.4 (*Poincaré's Theorem* or *Inclusion and Exclusion Formula*) For any events $A_1, A_2, \cdots, A_n$,

$$P\{A_1 \cup A_2 \cup \cdots \cup A_n\} = S_1 - S_2 + \cdots + (-1)^{n-1} S_n, \tag{1.4.15}$$

Table 1.4.1 Numerical examples of birthday problem.

r	Exact	Approx.
11	0.8589	0.8493
12	0.8330	0.8192
13	0.8056	0.7863
14	0.7769	0.7507
15	0.7471	0.7123
16	0.7164	0.6712
17	0.6850	0.6274
18	0.6531	0.5808
19	0.6209	0.5315
20	0.5886	0.4795
21	0.5563	0.4247
22	0.5243	0.3671
23	0.4927	0.3068
24	0.4617	0.2438
25	0.4313	0.1781
26	0.4018	0.1096
27	0.3731	0.0384
28	0.3455	-0.0356
29	0.3190	-0.1123
30	0.2937	-0.1918

where, for each $k = 1, 2, \cdots, n$, S_k is defined by

$$S_k = \sum_{i_1<i_2<\cdots<i_k} P\{A_{i_1} A_{i_2} \cdots A_{i_k}\} \qquad (i_1 \geq 1,\ i_k \leq n), \tag{1.4.16}$$

i.e., the summation is over all combinations of n events taken k at a time ($k = 1, 2, \cdots, n$).

Proof To compute the probability above, we should add the probabilities of all sample points contained in at least one of A_1, A_2, $\cdots$, A_n (i.e., S_1 has $\binom{n}{1}$ terms). However, we should subtract the probability of any sample point contained in at least two of A_1, A_2, $\cdots$, A_n (i.e., S_2 has $\binom{n}{2}$ terms), and so on. In general, the kth term S_k has $\binom{n}{k}$ terms. We intuitively prove the theorem. The reader should consult another textbook for a rigorous proof.

Special cases of $n = 2$ and 3 are the following:

$$P\{A_1 \cup A_2\} = P\{A_1\} + P\{A_2\} - P\{A_1 A_2\}, \tag{1.4.17}$$

and

$$P\{A_1 \cup A_2 \cup A_3\} = P\{A_1\} + P\{A_2\} + P\{A_3\}$$
$$-P\{A_1A_2\} - P\{A_2A_3\} - P\{A_1A_3\} + P\{A_1A_2A_3\}. \quad (1.4.18)$$

Thus, Poincaré's theorem is a generalization of Theorem 1.3.1 (iv) and Corollary 1.3.1.

Example 1.4.3 (*Matches* or *Coincidences*) There are n pairs of husbands and wives at a masquerade. Each husband proposes to be the dance partner of any wife randomly selected. What is the probability that at least one husband chooses his own wife? Let A_k be the event that the kth husband chooses his wife, where $k = 1, 2, \cdots, n$. Then $P\{A_1 \cup A_2 \cup \cdots \cup A_n\}$ is the probability that at least one husband chooses his wife, which is given in Eq.(1.4.15) of Theorem 1.4.4. It is clear that

$$P\{A_k\} = \frac{(n-1)!}{n!}, \qquad S_1 = \sum_{k=1}^{n} P\{A_k\} = \binom{n}{1}\frac{(n-1)!}{n!} = 1,$$

$$P\{A_iA_j\} = \frac{(n-2)!}{n!}, \qquad S_2 = \sum_{i<j} P\{A_iA_j\} = \binom{n}{2}\frac{(n-2)!}{n!} = \frac{1}{2!},$$

and, in general,

$$P\{A_{i_1}A_{i_2}\cdots A_{i_k}\} = \frac{(n-k)!}{n!}, \qquad S_k = \binom{n}{k}\frac{(n-k)!}{n!} = \frac{1}{k!}.$$

The last term of Eq.(1.4.15) is given by

$$S_n = P\{A_1A_2\cdots A_n\} = \frac{1}{n!}.$$

Finally, we have

$$P\{A_1 \cup A_2 \cup \cdots \cup A_n\} = 1 - \frac{1}{2!} + \frac{1}{3!} - \cdots + (-1)^{n-1}\frac{1}{n!}.$$

If n is large enough, we have the following approximation:

$$P\{A_1 \cup A_2 \cup \cdots \cup A_n\} \approx 1 - e^{-1} \doteq 0.63212,$$

since

$$e^{-1} = 1 - 1 + \frac{1}{2!} - \frac{1}{3!} + \frac{1}{4!} - \cdots$$

Table 1.4.2 shows the computational probabilities for $n = 3, \cdots, 10$. The probability of at least one match is approximated by $1 - e^{-1} \doteq 0.63212$ even if n

Table 1.4.2 Numerical examples of matches.

n	The probability
3	0.6667
4	0.6250
5	0.6333
6	0.6319
7	0.6321
8	0.6321
9	0.6321
10	0.6321

varies. This is the so-called *match* (*coincidence* or *rencontre*) problem and has several variations, given changes in the relevant terms.

We generalize from the binomial coefficient to the multinomial coefficient in the following:

Theorem 1.4.5 (*Multinomial Coefficient*) Let A contain n objects, and let $n_1, n_2, \cdots, n_r$ be non-negative integers such that $n_1 + n_2 + \cdots + n_r = n$. Then there exist

$$\binom{n}{n_1 n_2 \cdots n_r} = \frac{n!}{n_1! n_2! \cdots n_r!} \tag{1.4.19}$$

different ordered partitions of A of the ordered subsets $(A_1, A_2, \cdots, A_r)$, where A_k contains n_k objects, $k = 1, 2, \cdots, r$. Equation (1.4.19) is called the *multinomial coefficient.*

Proof First, we prove Eq.(1.4.19) for $r = 2$. The ordered subset sequence (A_1, A_2) forms a partition, and we choose n_1 objects from the subset A_1 and we choose $n_2 = n - n_1$ objects from the subset A_2 whose size is $n - n_1$. Thus

$$\binom{n}{n_1}\binom{n-n_1}{n-n_1} = \frac{n!}{n_1!(n-n_1)!} = \frac{n!}{n_1! n_2!}. \tag{1.4.20}$$

In general, we have

$$\binom{n}{n_1}\binom{n-n_1}{n_2}\binom{n-n_1-n_2}{n_3} \cdots \binom{n-n_1-\cdots-n_{k-2}}{n_{k-1}}\binom{n-n_1-\cdots-n_{k-1}}{n_k}$$

$$= \frac{n!}{n_1! n_2! \cdots n_k!}, \tag{1.4.21}$$

which is the multinomial coefficient in Eq.(1.4.19).

Example 1.4.4 (*Multinomial Expansion*)

$$(x+y+z)^n = \sum \binom{n}{n_1 n_2 n_3} x^{n_1} y^{n_2} z^{n_3},$$

where the summation is over all combinations of $n_1+n_2+n_3 = n$, n_1, n_2, $n_3 \geq 0$. For instance,

$$(x+y+z)^2 = x^2 + y^2 + z^2 + 2xy + 2yz + 2xz,$$

since

$$\binom{2}{200} = \binom{2}{020} = \binom{2}{002} = 1,$$

$$\binom{2}{110} = \binom{2}{011} = \binom{2}{101} = 2.$$

Example 1.4.5 (*Hypergeometric Distribution*) Consider an urn of n balls containing n_1 red balls and $n_2 = n - n_1$ black balls. We draw r balls without replacement at random. What is the probability P_k that k red balls are drawn (and, of course, $(r-k)$ black balls are drawn)? We draw k red balls out of n_1 red balls, i.e., $\binom{n_1}{k}$ different ways, and $(r-k)$ black balls out of $n - n_1$ black balls, i.e., $\binom{n-n_1}{r-k}$ different ways. Thus, since we have $\binom{n}{r}$ different ways of drawing r balls out of n balls, we have

$$P_k = \frac{\binom{n_1}{k}\binom{n-n_1}{r-k}}{\binom{n}{r}} = \frac{\binom{r}{k}\binom{n-r}{n_1-k}}{\binom{n}{n_1}},$$

where $\max(0, r-n_2) \leq k \leq \min(r, n_1)$. The last identity is derived from the identity in Eq.(1.4.11).

Example 1.4.6 (*Sampling Inspection*) In quality control, we often encounter a typical sampling inspection problem: Lots of size n are subject to sampling inspection. There are n balls (items) containing n_1 red balls (defective items) and $n_2 = n - n_1$ black balls (nondefective items). We draw r balls in which k red balls are counted. However, we cannot know the number n_1 of defective items in advance. The sampling inspection problem is how to estimate and test the number n_1 of defective items.

1.5 Problems 1

1.1 Let A, B, C denote any three events, Ω the sample space and ϕ the null event. Show the following set operations:

(i) $A \cup A = A$, $A \cap A = A$ (*Idempotent Laws*).

(ii) $(A \cup B) \cup C = A \cup (B \cup C)$, $(AB)C = A(BC)$ (*Associative Laws*).

(iii) $A \cup B = B \cup A$, $AB = BA$ (*Commutative Laws*).

(iv) $A \cup (BC) = (A \cup B)(A \cup C)$,
$A(B \cup C) = (AB) \cup (AC)$ (*Distributive Laws*).

(v) $A \cup \phi = A$, $A\Omega = A$,
$A \cup \Omega = \Omega$, $A\phi = \phi$ (*Identity Laws*).

(vi) $A \cup A^c = \Omega$, $AA^c = \phi$,
$(A^c)^c = A$, $\Omega^c = \phi$, $\phi^c = \Omega$ (*Complement Laws*).

(vii) $(A \cup B)^c = A^c B^c$, $(AB)^c = A^c \cup B^c$ (*De Morgan's Laws*).

1.2 Let A, B, C denote any three events. Find expressions for the events that of A, B, C:

(i) Only A occurs.

(ii) At least one occurs.

(iii) At least two occur.

(iv) All three events occur.

(v) One and no more occurs.

(vi) Two and no more occur.

(vii) None occur.

(viii) Not more than two occur.

1.3 Verify that, if two events A and B are independent, then A^c and B^c are independent.

1.4 Let B denote an event B such that $P\{B\} > 0$. Show that, for any event A, the conditional probability $P\{A \mid B\}$ satisfies the three axioms in Definition 1.3.1.

1.5 Let A and B denote two events with $P\{A\} = 1/3$, $P\{B\} = 1/4$ and $P\{A \cup B\} = 1/2$.

Find:

(i) $P\{A \mid B\}$.

(ii) $P\{B \mid A\}$.

(iii) $P\{A - B\}$.

(iv) $P\{B - A\}$.

1.6 If n, m, r are positive integers, verify

$$\binom{m}{0}\binom{n-m}{r} + \binom{m}{1}\binom{n-m}{r-1} + \cdots + \binom{m}{r}\binom{n-m}{0} = \binom{n}{r}.$$

1.7 (*Continuation*) Verify

$$\binom{n}{0}^2 + \binom{n}{1}^2 + \cdots + \binom{n}{n}^2 = \binom{2n}{n}.$$

1.8 (*Continuation*) Verify

$$\sum_{r=0}^{n} \frac{(2n)!}{(r!)^2[(n-r)!]^2} = \binom{2n}{n}^2.$$

1.9 Expand and simplify the following equations:

(i) $(3x^2 - 2y)^3$.

(ii) $(4x + 3y^2)^3$.

1.10 In how many ways can we choose a chairperson and three vice-chairpersons out of 50 persons?

1.11 Show that

$$\binom{-1}{r} = (-1)^r, \qquad \binom{-2}{r} = (-1)^r(r+1),$$

and verify that

$$(1+t)^{-1} = 1 - t + t^2 - t^3 + t^4 - \cdots,$$

$$(1+t)^{-2} = 1 - 2t + 3t^2 - 4t^3 + \cdots,$$

for $|\, t \,| < 1$.

1.12 (i) Enumerate all the possible outcomes of a random trial of placing three distinguishable balls (say a, b, c) into three cells. Take as an example $\{abc \mid - \mid -\}$, indicating three balls in the first cell. (ii) In how many ways do we have all possible outcomes?

1.13 (i) Enumerate all possible outcomes of a random trial of placing three indistinguishable balls into three cells. Take as an example $\{*** \mid - \mid -\}$, indicating three balls in the first cell. (ii) In how many ways do we have all possible outcomes?

1.14 (*Continuation*) In how many ways do we have all possible outcomes of a random trial of placing s distinguishable (and indistinguishable) balls into n cells, respectively?

1.15 (*Bridge*) Consider a deck of cards (see Example 1.2.1 (3)). What is the probability that a hand of thirteen cards consists of five spades, three clubs, three hearts, and two diamonds?

Chapter 2

Random Variables and Distributions

2.1 Introduction

Consider the random phenomenon of a queue at an Automated Teller's Machine (ATM) at a bank, where a queue is a waiting line. Potential customers arrive at the ATM to get cash or other services. Table 2.1.1 shows the observations of the moments of arrival of customers over the course of an hour. Two figures can be considered to describe the random phenomena in Table 2.1.1.

Figure 2.1.1 is a graph showing the cumulative number of arriving customers over the time axis, which is referred to as a "sample function" of a stochastic process.

We next consider the interarrival times X_1, X_2, $\cdots$, which can be calculated as the difference between the moments of arrival of two consecutive customers. Classifying these data into 12 classes, we obtain Table 2.1.2.

Figure 2.1.2 is a graph showing the number of data falling into each class, which is referred to as a "histogram." As will be described later, the probabilistic law of interarrival times can be described by an exponential distribution and the observed histogram fits quite well to the theoretical numbers of data falling into each class, as derived from an exponential distribution (see Problem 2.7).

Table 2.1.1 Observations of the moments of arrival of customers.

Number of customers	Inter arrival time	cumlative arrival time
1	30.86	30.86
2	95.87	126.73
3	69.02	195.75
4	45.29	241.04
5	40.24	281.28
6	1.46	282.74
7	12.86	295.60
8	32.68	328.28
9	62.93	391.21
10	32.94	424.15
11	95.19	519.34
12	20.33	539.67
13	45.83	585.50
14	28.50	614.00
15	115.04	729.04
16	9.71	738.75
17	24.37	763.12
18	12.36	775.48
19	73.81	849.29
20	5.84	855.13
21	11.70	866.83
22	3.04	869.87
23	64.47	933.34
24	90.23	1023.57
25	47.68	1071.25
26	30.63	1101.88
27	10.51	1112.39
28	16.55	1128.94
29	19.89	1148.83
30	4.81	1153.64
31	14.88	1168.52
32	101.79	1270.31
33	22.57	1292.88
34	39.79	1332.67
35	22.50	1355.17
36	9.28	1364.45
37	88.30	1452.75
38	58.22	1510.97
39	50.92	1561.89
40	18.40	1580.29
41	15.93	1596.22
42	33.14	1629.36
43	31.48	1660.84
44	46.75	1707.59
45	13.75	1721.34
46	101.01	1822.35
47	.15	1822.50
48	2.51	1825.01
49	65.65	1890.66
50	4.03	1894.69
51	22.50	1917.19
52	26.54	1943.73
53	87.43	2031.16
54	9.42	2040.58
55	89.15	2129.73
56	33.97	2163.70
57	1.32	2165.02
58	53.86	2218.88
59	47.13	2266.01
60	1.77	2267.78
61	51.35	2319.13
62	.94	2320.07
63	80.51	2400.58
64	10.31	2410.89
65	103.75	2514.64
66	56.54	2571.18
67	46.33	2617.51
68	33.15	2650.66
69	35.67	2686.33
70	27.45	2713.78
71	6.80	2720.58
72	4.21	2724.79
73	6.61	2731.40
74	6.89	2738.29
75	29.11	2767.40
76	209.16	2976.56
77	1.29	2977.85
78	29.09	3006.94
79	54.27	3061.21
80	.59	3061.80
81	5.66	3067.46
82	49.01	3116.47
83	1.67	3118.14
84	1.47	3119.61
85	16.68	3136.29
86	51.00	3187.29
87	17.57	3204.86
88	31.53	3236.39
89	1.29	3237.68
90	97.44	3335.12
91	79.57	3414.69
92	5.90	3420.59
93	32.93	3453.52
94	27.29	3480.81
95	25.93	3506.64
96	.66	3507.30
97	58.78	3566.08
98	1.57	3567.65
99	15.57	3584.22

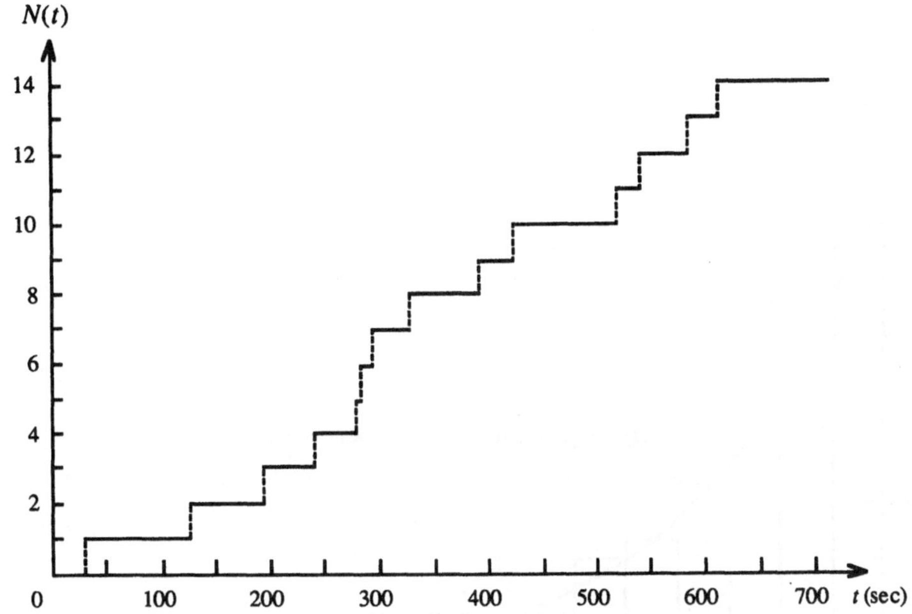

Fig. 2.1.1 The cumulative number of arriving customers over the time axis.

Table 2.1.2 The number of data falling into each class.

t	The number of arriving customers
$0 \leq t < 10$	27
$10 \leq t < 20$	14
$20 \leq t < 30$	12
$30 \leq t < 40$	12
$40 \leq t < 50$	8
$50 \leq t < 60$	7
$60 \leq t < 70$	4
$70 \leq t < 80$	2
$80 \leq t < 90$	4
$90 \leq t < 100$	4
$100 \leq t < 110$	3
$110 \leq t$	2

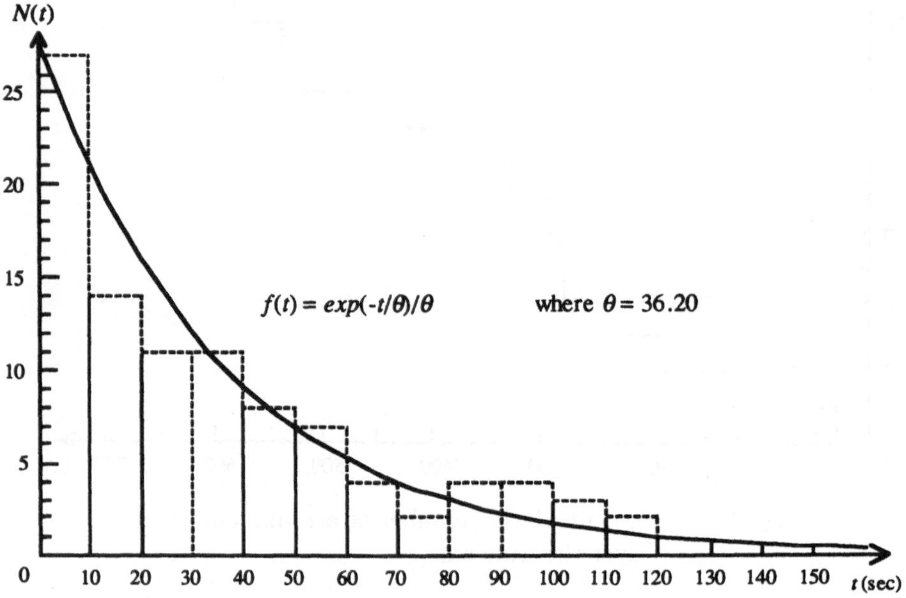

Fig. 2.1.2 A histogram of Table 2.1.2 and the corresponding exponential density curve.

The main purpose of this book is to introduce basic stochastic processes and apply them to real-world problems. A stochastic process can be described by the laws of probability at any point of time. That is, specifying any time, we can describe the random phenomenon by the law of probability. A queue at the ATM just cited is a typical example of a stochastic process.

Section 2.2 deals with random variables and probability distributions. Sections 2.3 and 2.4 discuss common distributions which will frequently be used in later discussions. Finally, Sections 2.5 and 2.6 deal with multivariate distributions and limit theorems, respectively.

2.2 Random Variables and Distributions

It is plausible and convenient to consider a real-valued function whose domain is a sample space Ω. Consider a real-valued function f from the sample space Ω into the real space R, i.e., a function f that assigns a real value to each sample point ω. In particular, we denote the *image* $f(A)$ that assigns a subset of R to the event A. We also denote the *preimage* $f^{-1}(B)$ that assigns a subset of $\omega \in \Omega$ such that $f(\omega) \in B \subset R$.

Definition 2.2.1 A real-valued function X whose domain is the sample space Ω is called a *random variable.*

In other words, a random variable X in a sample space Ω is a function from Ω into the real space R such that the preimage of every interval of R is an event of Ω. In this section, we restrict ourselves to a one-dimensional real space R.

If we specify the interval $(a, b]$ $(a < b)$ of R, we can calculate the following probability

$$P\{a < X \leq b\} = P\{\omega \in \Omega : X(\omega) \in (a, b]\}, \tag{2.2.1}$$

from Definition 2.2.1.

Definition 2.2.2 The *probability distribution* or simply *distribution* of the random variable X is defined for any real number x by

$$F_X(x) = P\{X \leq x\} = P\{\omega \in \Omega : X(\omega) \in (-\infty, x]\}, \tag{2.2.2}$$

which is the probability that the random variable X is less than or equal to x.

Theorem 2.2.1 The distribution $F_X(x)$ of the random variable X is satisfied by the following properties:

(i) $F_X(x)$ is monotonically increasing and $0 \leq F_X(x) \leq 1$.

(ii) $F_X(x)$ is right continuous, i.e., $\lim_{h \downarrow 0} F_X(x+h) = F_X(x)$.

(iii) $\lim_{x \to \infty} F_X(x) = 1$ and $\lim_{x \to -\infty} F_X(x) = 0$.

A random variable X is said to be *discrete* if its set of all possible values is countable or denumerable. For discrete random variables, the *probability mass function* $p_X(x_i)$ is defined by

$$p_X(x_i) = P\{X = x_i\} = P\{\omega \in \Omega : X(\omega) = x_i\}, \tag{2.2.3}$$

such that $p_X(x_i) \geq 0$ for $i = 1, 2, \cdots$ and $\sum_{i=1}^{\infty} p_X(x_i) = 1$. Then, the distribution of the discrete random variable X is given by

$$\begin{aligned} F_X(x) &= P\{X \leq x\} \\ &= P\{\omega \in \Omega : X(\omega) \in (-\infty, x]\} \\ &= \sum_{y \leq x} p_X(y). \end{aligned} \tag{2.2.4}$$

A random variable X is said to be *continuous* if there exists a *density* $f_X(x)$ such that

$$F_X(x) = P\{X \leq x\} = \int_{-\infty}^{x} f_X(y)\, dy. \tag{2.2.5}$$

It is clear from the above equation that

$$f_X(x) = \frac{dF_X(x)}{dx}. \tag{2.2.6}$$

For a *discrete* random variable X, the *expectation* or *mean* of the random variable X is defined by

$$E[X] = \sum_{i=1}^{\infty} x_i \, p_X(x_i), \tag{2.2.7}$$

provided the above sum exists. The expectation is the weighted sum of X by the weight $p_X(x_i)$, and is referred to as the center of gravity in mechanics.

For a *continuous* random variable X, the *expectation* or *mean* of the random variable X is defined by

$$E[X] = \int_{-\infty}^{\infty} x f_X(x) \, dx, \tag{2.2.8}$$

provided the above integral exists.

It is quite cumbersome to distinguish random variables as discrete or continuous. To overcome this, we introduce the concept of Stieltjes integrals. Combining Eqs.(2.2.7) and (2.2.8), we denote the expectation of the random variable X by

$$E[X] = \int_{-\infty}^{\infty} x dF(x) = \begin{cases} \sum_{i=1}^{\infty} x_i \, p_X(x_i) & \text{if } x \text{ is discrete} \\ \int_{-\infty}^{\infty} x f_X(x) \, dx & \text{if } x \text{ is continuous,} \end{cases} \tag{2.2.9}$$

which is expressed in terms of *Stieltjes integrals*. We never intend to present arguments for Stieltjes integrals in general. In the remainder of this book, we are just interested in discrete or continuous random variables. The expectation in Eq.(2.2.9) can be interpreted by Eq.(2.2.7) if X is discrete, or by Eq.(2.2.8) if X is continuous. Throughout this book, we use Stieltjes integrals which are assumed to be a sum such as Eq.(2.2.7) if X is discrete, or an integral such as Eq.(2.2.8) if X is continuous.

If X is a random variable and g is a real-valued function whose domain is the real space, then $g(X)$ is also a random variable. The expectation of $g(X)$ is given by

$$E[g(X)] = \int_{-\infty}^{\infty} g(x) \, dF_X(x), \tag{2.2.10}$$

provided the above integral exists.

The expectation of X^2 is given by

$$E[X^2] = \int_{-\infty}^{\infty} x^2 dF_X(x), \tag{2.2.11}$$

provided the above integral exists. Equation (2.2.11) is called the *second moment* of X about the origin. The *variance* of the random variable X is given by

$$Var(X) = E[(X - E[X])^2]$$

$$
\begin{aligned}
&= \int_{-\infty}^{\infty} (x - E[X])^2 dF_X(x) \\
&= \int_{-\infty}^{\infty} \left(x^2 - 2xE[X] + E[X]^2\right) dF_X(x) \\
&= \int_{-\infty}^{\infty} x^2 dF_X(x) - 2E[X] \int_{-\infty}^{\infty} x\, dF_X(x) + E[X]^2 \\
&= \int_{-\infty}^{\infty} x^2 dF_X(x) - E[X]^2 \\
&= E[X^2] - E[X]^2,
\end{aligned} \tag{2.2.12}
$$

which is referred to as the second moment of the random variable X about its mean $E[X]$. The *standard deviation* of the random variable X is defined by the positive root of the variance, $\sqrt{Var(X)}$. The *standardized random variable* Y corresponding to X is defined by

$$
Y = \frac{X - E[X]}{\sqrt{Var(X)}}, \tag{2.2.13}
$$

where $E[Y] = 0$ and $Var(Y) = 1$.

In general, the nth moment of the random variable X about the *origin* is defined by

$$
E[X^n] = \int_{-\infty}^{\infty} x^n dF_X(x) \qquad (n = 1, 2, \cdots), \tag{2.2.14}
$$

provided the above integral exists. Similarly, the nth moment of the random variable X about its mean is defined by

$$
E[(X - E[X])^n] = \int_{-\infty}^{\infty} (x - E[X])^n dF_X(x) \qquad (n = 1, 2, \cdots), \tag{2.2.15}
$$

provided the above integral exists. It is sometimes convenient to introduce the dimensionless moments about its mean:

$$
a_n = \frac{E[(X - E[X])^n]}{\left(\sqrt{Var(X)}\right)^n} \qquad (n = 1, 2, 3, \cdots). \tag{2.2.16}
$$

Noting $E[X - E[X]] = E[X] - E[X] = 0$ and $Var(X) = E[(X - E[X])^2]$, we have that $a_1 = 0$ and $a_2 = 1$. Table 2.2.1 shows the dimensionless moments a_n about their respective means for $n = 1, 2, 3, 4$.

The *characteristic function* of the random variable X is defined by

$$
\varphi_X(u) = E[e^{iuX}] = \int_{-\infty}^{\infty} e^{iux} dF_X(x), \tag{2.2.17}
$$

provided the above integral exists, where $i = \sqrt{-1}$ is an imaginary unit.

Theorem 2.2.2 The probability distribution is in one-to-one correspondence with its characteristic function.

Table 2.2.1 The dimensionless moments a_n ($n = 1, 2, 3, 4$).

n	a_n
1	0
2	1
3	$\frac{E[X^3] - 3E[X^2]E[X] + 2E[X]^3}{(\sqrt{Var(X)})^3}$
4	$\frac{E[X^4] - 4E[X^3]E[X] + 6E[X^2]E[X]^2 - 3E[X]^4}{(\sqrt{Var(X)})^4}$

a_3: moment coefficient of skewness.
a_4: moment coefficient of kurtosis.

Let X and Y be random variables for which the characteristic functions $\varphi_X(u)$ and $\varphi_Y(u)$ exist. Then $F_X(\cdot) = F_Y(\cdot)$ if and only if $\varphi_X(u) = \varphi_Y(u)$.

The nth moment of the random variable X about the origin can be easily derived from its characteristic function as follows:

$$E[X^n] = i^{-n} \left. \frac{d^n \varphi_X(u)}{du^n} \right|_{u=0}. \tag{2.2.18}$$

Equation (2.2.18) can be easily verified by the Taylor expansion of the characteristic function:

$$\varphi_X(u) = \sum_{n=0}^{\infty} \frac{(iu)^n}{n!} \int_{-\infty}^{\infty} x^n dF_X(x), \tag{2.2.19}$$

and n times differentiations.

In statistics, instead of the characteristic function, the *moment generating function* of the random variable X is defined by

$$M_X(\theta) = \int_{-\infty}^{\infty} e^{\theta x} dF_X(x), \tag{2.2.20}$$

provided the above integral exists.

In stochastic processes, the *Laplace-Stieltjes* transforms can be adopted for the non-negative random variables, since the stochastic processes are concerned with the non-negative random variables that represent the real time $t \geq 0$. The Laplace-Stieltjes transform of the non-negative random variable X is defined by

$$F_X^*(s) = \int_0^{\infty} e^{-sx} dF_X(x), \tag{2.2.21}$$

provided the above integral exists, where $\Re(s) > 0$.

Table 2.2.2 shows the formulas for the characteristic function, moment generating function and Laplace-Stieltjes transform.

Table 2.2.2 Definition, Theorem 2.2.1 and moments of the integral transforms.

Integral Transform	Definition	Theorem 2.2.1	Moments
Characteristic Function	$\varphi_X(u) = \int_{-\infty}^{\infty} e^{iux} dF_X(x)$	$F_X(\cdot) = F_Y(\cdot)$ $\Updownarrow$ $\varphi_X(u) = \varphi_Y(u)$	$E[X^n] = i^{-n} \left. \frac{d^n \varphi_X(u)}{du^n} \right\|_{u=0}$
Moment Generating Function	$M_X(\theta) = \int_{-\infty}^{\infty} e^{\theta x} dF_X(x)$	$F_X(\cdot) = F_Y(\cdot)$ $\Updownarrow$ $M_X(\theta) = M_Y(\theta)$	$E[X^n] = \left. \frac{d^n M_X(\theta)}{d\theta^n} \right\|_{\theta=0}$
Laplace-Stieltjes Transform $(x \geq 0)$	$F_X^*(s) = \int_{0}^{\infty} e^{-sx} dF_X(x)$ $(Re(s)>0)$	$F_X(\cdot) = F_Y(\cdot)$ $\Updownarrow$ $F_X^*(s) = F_Y^*(s)$	$E[X^n] = (-1)^n \left. \frac{d^n F_X^*(s)}{ds^n} \right\|_{s=0}$

2.3 Discrete Distributions

In this section we introduce common probability distributions that are well-known in probability theory. In particular, we present six *discrete* random variables with their probability mass functions. In the next section, we present seven *continuous* random variables with their densities (or distributions). There are many relationships among these distributions. We cite several relationships among these distributions in this and following sections.

In this section we introduce six common discrete distributions. Each name, symbol, probability mass function, mean, variance, and characteristic function (if it exists) is described.

(i) Uniform Distribution $X \sim U(C + L, C + NL)$

Consider a fair die casting. The random variable is designated by the number appearing i (i = 1, 2, 3, 4, 5, 6). Then the probability mass function of the number appearing i is $P\{X = i\} = 1/6$ (i = 1, 2, 3, 4, 5, 6).

Generalizing this random trial, we introduce the probability mass function of the discrete uniform random variable X for constants C and L $(L > 0)$:

$$p_X(C + xL) = P\{X = C + xL\} = \frac{1}{N} \qquad (x = 1, 2, \cdots, N), \quad (2.3.1)$$

which is referred to as the probability mass function of the *discrete uniform distribution* and is denoted by $X \sim U(C+L, C+NL)$, where C, L $(L>0)$, and N are parameters. The expectation can be calculated by

$$E[X] = \sum_{j=1}^{N}(C+jL)p_X(C+jL) = C + \frac{(N+1)L}{2}, \tag{2.3.2}$$

and the second moment of the random variable X about the origin can be calculated by

$$\begin{aligned} E[X^2] &= \sum_{j=1}^{N}(C+jL)^2 p_X(C+jL) \\ &= C^2 + (N+1)LC + \frac{(N+1)(2N+1)L^2}{6}, \end{aligned} \tag{2.3.3}$$

where we use a formula of sum $\sum_{j=1}^{n} j^2$ (see Problem 2.1). From Eq.(2.3.3), we have the variance of the random variance X:

$$Var(X) = E[X^2] - E[X]^2 = \frac{(N^2-1)L^2}{12}. \tag{2.3.4}$$

The characteristic function of the discrete random variable X can be calculated by

$$\begin{aligned} \varphi_X(u) &= E[e^{iuX}] \\ &= \sum_{j=1}^{N} \frac{e^{iu(C+jL)}}{N} = \frac{e^{iu(C+L)}\left(1-e^{iuLN}\right)}{N\left(1-e^{iuL}\right)}. \end{aligned} \tag{2.3.5}$$

Example 2.3.1 The random digits (i.e., the discrete random numbers) are uniformly distributed on $x = 0, 1, 2, \cdots, 9$. Then, assuming that $C=-1$, $L=1$ and $N=10$, we have $E[X]=4.5$ and $Var(X)=33/4$.

In the following, we define the discrete random variable X assuming $x = 0, 1, 2, 3, \cdots$ unless otherwise specified. Of course, we can easily generalize $X = C + xL$ $(x = 0, 1, 2, \cdots)$ by introducing the appropriate constants C and L $(L > 0)$, which has just been shown above.

(ii) Bernoulli Distribution $X \sim Bernoulli(p)$

Assume that we perform a trial whose outcome can be classified as either "success" or "failure," with the respective probabilities p $(0<p<1)$ and $q = 1-p$. Here we assume that the trials are mutually independent. Such a trial is called a *Bernoulli trial.*

Let us consider the *Bernoulli distribution.* Define the random variable X, assuming 1 if the event (i.e., success) takes place, or 0 if the event does not take place. We often use the *indicator* of an event A as

$$\mathbf{1}(A) = \begin{cases} 1 & \text{if A takes place} \\ 0 & \text{if A does not take place.} \end{cases} \tag{2.3.6}$$

Then $\mathbf{1}(A)$ is a *Bernoulli* random variable with parameter $p = P\{A\}$. It is quite easy to show that

$$E[X] = 1 \cdot p + 0 \cdot (1-p) = p, \tag{2.3.7}$$

$$E[X^2] = 1^2 \cdot p + 0^2 \cdot (1-p) = p, \tag{2.3.8}$$

$$Var(X) = E[X^2] - E[X]^2 = p - p^2 = p(1-p) = pq. \tag{2.3.9}$$

The Bernoulli distribution is one of the simplest discrete distributions and is basic to the following three distributions which presuppose the Bernoulli trials.

(iii) Binomial Distribution $X \sim B(n, p)$

Consider the Bernoulli trials, where the number of trials is fixed, say $n > 0$. Then the probability of x successes is given by

$$p_X(x) = \binom{n}{x} p^x q^{n-x} \qquad (x = 0, 1, 2, \cdots, n), \tag{2.3.10}$$

which is the probability mass function of the *binomial distribution* and is denoted by $X \sim B(n, p)$, where n (n : a positive integer) and p ($0 < p < 1$) are parameters. Figure 2.3.1 shows the probability mass functions of the binomial distribution $X \sim B(n, p)$, where $n = 10$, $p = 0.2, 0.5, 0.9$.

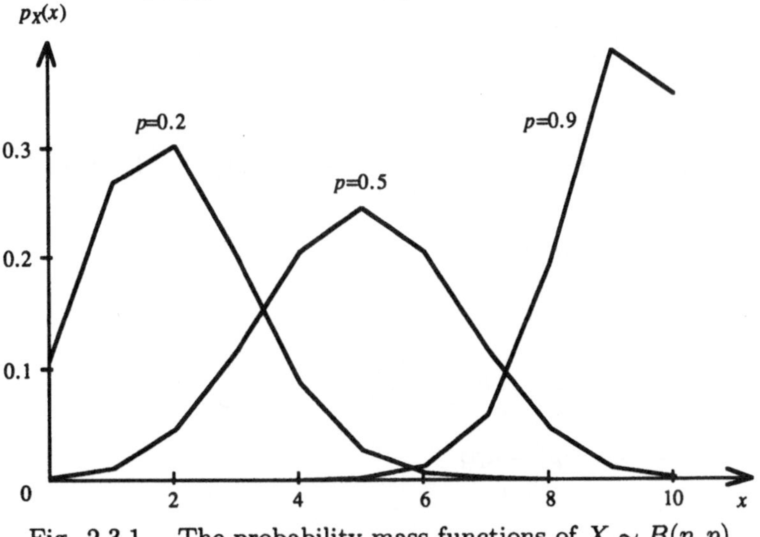

Fig. 2.3.1 The probability mass functions of $X \sim B(n,p)$, where $n = 10; p = 0.2, 0.5, 0.9$.

The mean, variance, and characteristic function are given by

$$E[X] = \sum_{x=1}^{n} x \binom{n}{x} p^x q^{n-x}$$

$$= \sum_{x=1}^{n} x \frac{n!}{x!(n-x)!} p^x q^{n-x}$$

$$= np \sum_{x=1}^{n} \binom{n-1}{x-1} p^{x-1} q^{n-x}$$

$$= np, \tag{2.3.11}$$

$$Var(X) = E[X^2] - E[X]^2 = npq, \tag{2.3.12}$$

and

$$\varphi_X(u) = \sum_{x=0}^{n} e^{iux} \binom{n}{x} p^x q^{n-x} = \left(pe^{iu} + q\right)^n, \tag{2.3.13}$$

respectively.

For Eqs.(2.3.11) and (2.3.12), we have directly calculated the mean and variance of the random variable X. Applying the formula (2.2.18), we have

$$E[X] = i^{-1} \frac{d\varphi_X(u)}{du}\bigg|_{u=0} = npe^{iu} \left(pe^{iu} + q\right)^{n-1}\bigg|_{u=0} = np, \tag{2.3.14}$$

and

$$E[X^2] = i^{-2} \frac{d^2\varphi_X(u)}{du^2}\bigg|_{u=0}$$

$$= npe^{iu} \left(pe^{iu} + q\right)^{n-1}\bigg|_{u=0} + n(n-1)\left(pe^{iu}\right)^2 \left(pe^{iu} + q\right)^{n-2}\bigg|_{u=0}$$

$$= np + n(n-1)p^2, \tag{2.3.15}$$

from which, we have

$$Var(X) = np + n(n-1)p^2 - n^2p^2$$

$$= np(1-p) = npq. \tag{2.3.16}$$

(iv) Geometric Distribution $X \sim GEO(p)$

Consider the Bernoulli trials. The relation of probability between necessary trials and first success is given by

$$p_X(x) = pq^{x-1} \qquad (x = 1, 2, \cdots), \tag{2.3.17}$$

which is the probability mass function of the *geometric distribution* and is denoted by $X \sim GEO(p)$, where p $(0 < p < 1)$ is a parameter. Figure 2.3.2 shows the probability mass functions of the geometric distribution $X \sim GEO(p)$, where $p = 0.2$, 0.5, 0.9.

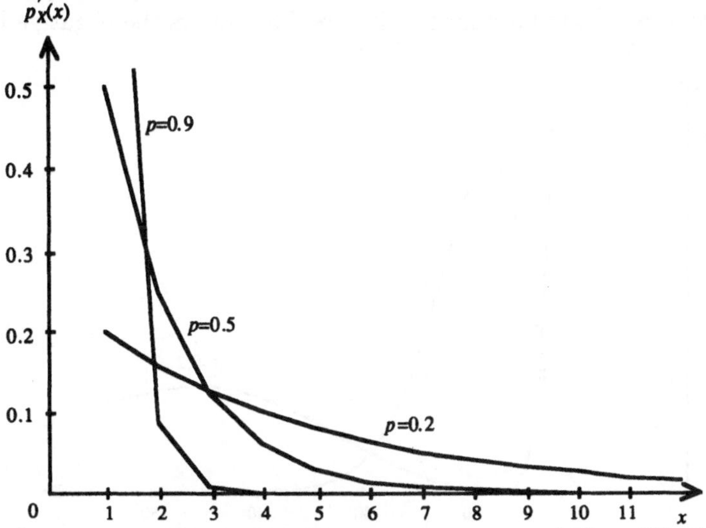

Fig. 2.3.2 The probability mass functions of $X \sim GEO(p)$, where $p = 0.2, 0.5, 0.9$.

The expectation can be calculated by

$$E[X] = \sum_{x=1}^{\infty} xpq^{x-1} = p\left(1 + 2q + 3q^2 + \cdots\right) = \frac{1}{p}. \tag{2.3.18}$$

The variance and characteristic function are given by

$$Var(X) = \sum_{x=1}^{\infty} x^2pq^{x-1} - E[X]^2 = \frac{q}{p^2}, \tag{2.3.19}$$

and

$$\varphi_X(u) = \sum_{x=1}^{\infty} e^{iux}pq^{x-1} = \frac{pe^{iu}}{1 - qe^{iu}}. \tag{2.3.20}$$

(v) Negative Binomial Distribution $X \sim NB(p, r)$

Consider the Bernoulli trials. The probability of necessary trials to first r successes ($r \geq 1$) is given by

$$\begin{aligned} p_X(x) &= \binom{x-1}{x-r} p^r q^{x-r} \\ &= \frac{(x-1)(x-2)\cdots(r+1)r}{(x-r)!} p^r q^{x-r} \\ &= \binom{-r}{x-r} p^r (-q)^{x-r} \qquad (x = r,\, r+1,\, \cdots), \end{aligned} \tag{2.3.21}$$

which is the probability mass function of the *negative binomial distribution* or *Pascal distribution* and is denoted by $X \sim NB(p, r)$, where p $(0 < p < 1)$ and r (r : a positive integer) are parameters. Figure 2.3.3 shows the negative binomial distribution $X \sim NB(p, r)$, where $p = 0.2,\ 0.5,\ 0.9;\ r = 10$.

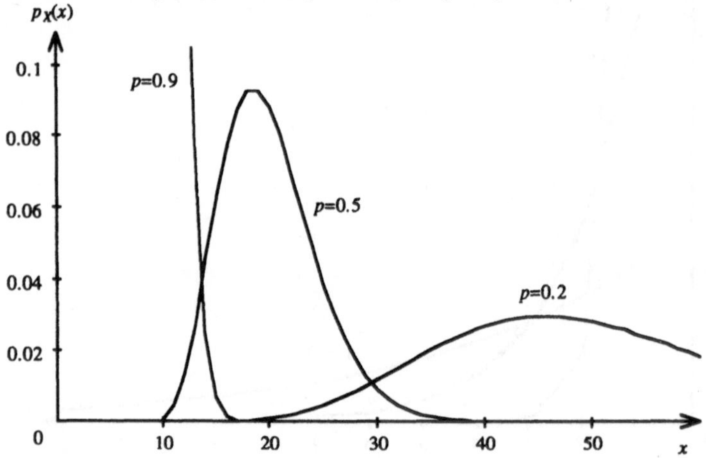

Fig. 2.3.3 The probability mass functions of $X \sim NB(p, r)$, where $p = 0.2, 0.5, 0.9; r = 10$.

Note that the negative binomial coefficient is defined by

$$\begin{aligned}\binom{-r}{x-r} &= \frac{(-r)(-r-1)\cdots(-r-x+r+1)}{(x-r)!} \\ &= \frac{(x-1)(x-2)\cdots(r+1)r}{(x-r)!}(-1)^{x-r} \\ &= \binom{x-1}{x-r}(-1)^{x-r}. \qquad (2.3.22)\end{aligned}$$

The mean, variance, and characteristic function are given by

$$E[X] = \frac{r}{p}, \qquad (2.3.23)$$

$$Var(X) = \frac{rq}{p^2}, \qquad (2.3.24)$$

and

$$\varphi_X(u) = \left(\frac{pe^{iu}}{1-qe^{iu}}\right)^r, \qquad (2.3.25)$$

respectively. Note that $NB(p, 1) = GEO(p)$, which will be interpreted in the following example.

Example 2.3.2 Let us consider the Bernoulli trials with probability p of success. The negative binomial distribution is concerned with the probability of necessary trials of failure to first r successes ($r \geq 1$). That is, consider the probability of necessary trials of failure for $(x-1)$ trials and that the final trial is successful:

$$\begin{aligned} &\binom{x-1}{x-r} p^{x-1-(x-r)} q^{x-r} \cdot p \\ &= \binom{x-1}{x-r} p^r q^{x-r}, \end{aligned}$$

which is the probability mass function of the negative binomial distribution $X \sim NB(p, r)$. Of course, if $X_1, X_2, \cdots, X_r$ are independent and identically distributed random variables with $X_i \sim GEO(p)$, the characteristic function of $S_r = X_1 + X_2 + \cdots + X_r$ is given by

$$\varphi_{S_r}(u) = \left(\frac{pe^{iu}}{1 - qe^{iu}}\right)^r,$$

which is that of the negative binomial distribution $S_r \sim NB(p, r)$ (see Theorem 2.2.2). In particular, if $r = 1$, then $X \sim NB(p, 1) = GEO(p)$, i.e., necessary trials to first success are distributed geometrically.

(vi) Poisson Distribution $X \sim POI(\lambda)$

The probability mass function of the *Poisson distribution* is given by

$$p_X(x) = \frac{e^{-\lambda}\lambda^x}{x!} \qquad (x = 0, 1, 2, \cdots), \tag{2.3.26}$$

and is denoted by $X \sim POI(\lambda)$, where λ ($\lambda > 0$) is a parameter. Figure 2.3.4 shows the probability mass functions of the Poisson distribution $X \sim POI(\lambda)$, where $\lambda = 0.5, 1, 2, 5$.

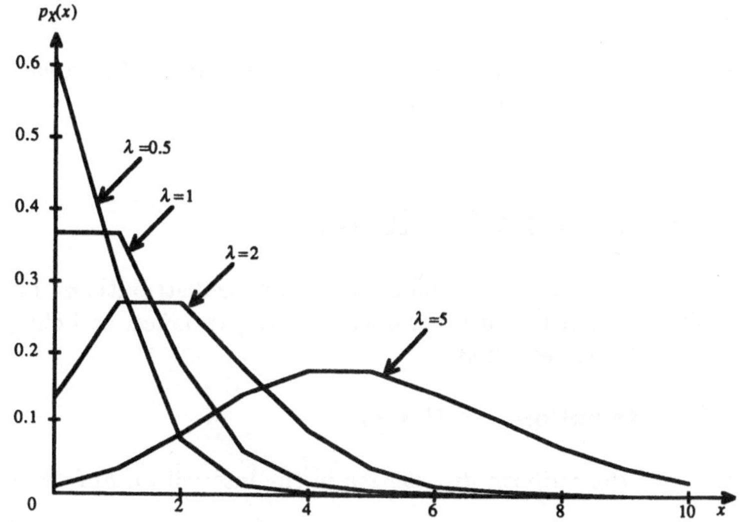

Fig. 2.3.4 The probability mass functions of the Poisson distribution $X \sim POI(\lambda)$, where $\lambda = 0.5, 1, 2, 5$.

The mean, variance, and characteristic function are given by

$$E[X] = \lambda \sum_{x=0}^{\infty} x e^{-\lambda} \frac{\lambda^{x-1}}{x!} = \lambda \sum_{x=0}^{\infty} e^{-\lambda} \frac{\lambda^{x}}{x!} = \lambda, \tag{2.3.27}$$

$$Var(X) = E[X^2] - E[X]^2 = \lambda, \tag{2.3.28}$$

and

$$\varphi_X(u) = \sum_{x=0}^{\infty} e^{iux} e^{-\lambda} \frac{\lambda^x}{x!} = e^{-\lambda} \cdot e^{\lambda e^{iu}} = \exp\left[\lambda\left(e^{iu} - 1\right)\right], \tag{2.3.29}$$

respectively.

Example 2.3.3 Consider the binomial distribution $X \sim B(n, p)$. Fixing $np = \lambda$ and letting $n \to \infty$, we have

$$\begin{aligned}
&\lim_{n\to\infty} p_X(x) \\
&= \lim_{n\to\infty} \frac{n(n-1)\cdots(n-x+1)}{x!} p^x (1-p)^{n-x} \\
&= \lim_{n\to\infty} (np)^x \frac{\left(1-\frac{1}{n}\right)\left(1-\frac{2}{n}\right)\cdots\left(1-\frac{x-1}{n}\right)}{x!} (1-p)^{n-x} \\
&= \lim_{n\to\infty} \lambda^x \frac{\left(1-\frac{1}{n}\right)\left(1-\frac{2}{n}\right)\cdots\left(1-\frac{x-1}{n}\right)}{x!} \left(1-\frac{\lambda}{n}\right)^{n-x} \\
&= \frac{\lambda^x}{x!} e^{-\lambda}
\end{aligned}$$

where $\left(1-\frac{\lambda}{n}\right)^{n-x} \to e^{-\lambda}$ as $n \to \infty$. The right-hand side of the equation above is the probability mass function of the Poisson distribution.

2.4 Continuous Distributions

In this section we introduce seven common continuous distributions. Each name, symbol, density (or distribution), parameter, mean, variance, and characteristic function (if it exists) is described.

(i) Uniform Distribution $X \sim U(a, b)$

The density of the *uniform distribution* in the interval (a, b) $(a < b)$ is given by

$$f_X(x) = \frac{1}{b-a} \qquad (a < x < b). \tag{2.4.1}$$

The corresponding distribution is given by

$$F_X(x) = \begin{cases} 0 & (x \leq a) \\ \frac{x-a}{b-a} & (a < x < b) \\ 1 & (x \geq b). \end{cases} \tag{2.4.2}$$

The uniform distribution in the interval (a, b) is denoted by $X \sim U(a, b)$, where a and b $(a < b)$ are parameters. The mean, variance, and characteristic function are given by

$$\begin{aligned} E[X] &= \int_a^b \frac{x}{b-a} dx \\ &= \frac{1}{b-a} \cdot \frac{x^2}{2}\Big|_a^b = \frac{1}{b-a} \cdot \frac{b^2 - a^2}{2} = \frac{a+b}{2}, \end{aligned} \tag{2.4.3}$$

$$\begin{aligned} E[X^2] &= \int_a^b \frac{x^2}{b-a} dx \\ &= \frac{1}{b-a} \cdot \frac{x^3}{3}\Big|_a^b \\ &= \frac{1}{b-a} \cdot \frac{b^3 - a^3}{3} \\ &= \frac{a^2 + ab + b^2}{3}, \end{aligned} \tag{2.4.4}$$

and

$$\begin{aligned} Var(X) &= E[X^2] - E[X]^2 \\ &= \frac{a^2 + ab + b^2}{3} - \left(\frac{a+b}{2}\right)^2 \\ &= \frac{(b-a)^2}{12}, \end{aligned} \tag{2.4.5}$$

$$\varphi_X(u) = \int_a^b \frac{e^{iux}}{b-a} dx = \frac{1}{iu(b-a)} e^{iux}\Big|_a^b = \frac{e^{iub} - e^{iua}}{iu(b-a)}, \tag{2.4.6}$$

respectively.

In particular, the uniform distribution $X \sim U(0, 1)$ is called the *standard uniform distribution* and corresponds to the continuous random number in the interval [0, 1]. The continuous random number can be easily generated by microcomputers (e.g., RND in BASIC Command). Figure 2.4.1 shows the density and distribution of the uniform distribution $X \sim U(0,\ 1)$.

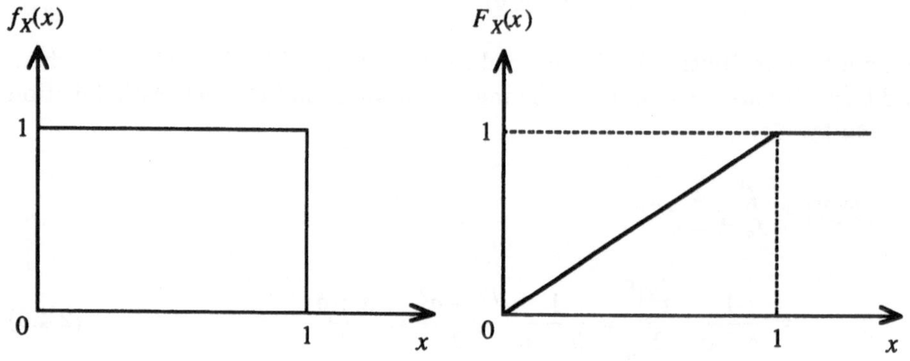

(a) Density (b) Distribution

Fig. 2.4.1 The density and distribution of $X \sim U(0, 1)$.

(ii) Exponential Distribution $X \sim EXP(\lambda)$

The density of the *exponential distribution* is given by

$$f_X(x) = \begin{cases} 0 & (x < 0) \\ \lambda e^{-\lambda x} & (x \geq 0), \end{cases} \tag{2.4.7}$$

and is denoted by $X \sim EXP(\lambda)$, where λ $(\lambda > 0)$ is a parameter. The corresponding distribution is given by

$$F_X(x) = \begin{cases} 0 & (x < 0) \\ 1 - e^{-\lambda x} & (x \geq 0). \end{cases} \tag{2.4.8}$$

Figure 2.4.2 shows the density and distribution of the exponential distribution

$X \sim EXP(\lambda)$, where $\lambda = 1$.

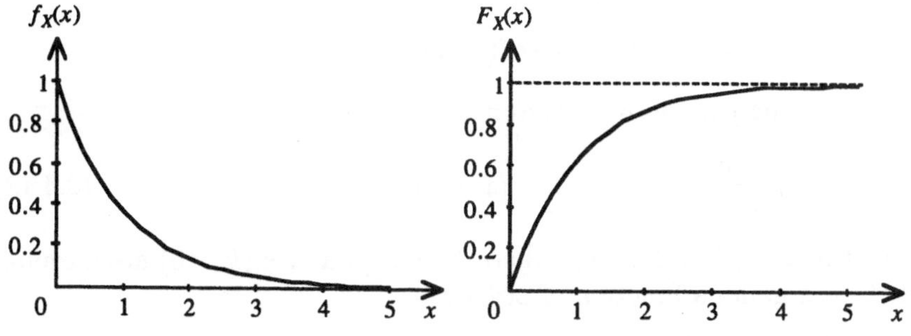

Fig. 2.4.2 The density and distribution of $X \sim EXP(1)$.

The mean can be calculated by

$$\begin{aligned} E[X] &= \int_0^\infty x\lambda e^{-\lambda x}dx \\ &= -xe^{-\lambda x}\Big|_0^\infty + \int_0^\infty e^{-\lambda x}dx \\ &= -\frac{1}{\lambda}e^{-\lambda x}\Big|_0^\infty \\ &= \frac{1}{\lambda}, \end{aligned} \tag{2.4.9}$$

and the variance can be calculated by

$$Var(X) = \int_0^\infty x^2\lambda e^{-\lambda x}dx - E[X]^2 = \frac{1}{\lambda^2}. \tag{2.4.10}$$

Since the exponential random variable is non-negative, it is convenient to calculate the Laplace-Stieltjes transform:

$$F_X^*(s) = \int_0^\infty e^{-sx}\lambda e^{-\lambda x}dx = -\frac{\lambda}{s+\lambda}e^{-(s+\lambda)x}\Big|_0^\infty = \frac{\lambda}{s+\lambda}. \tag{2.4.11}$$

Example 2.4.1 In Fig. 2.1.2, we have shown that the observed histogram fits the exponential density. We show that the theoretical numbers of data falling into each class can be calculated by

$$P\{a < X \le b\} = F_X(b) - F_X(a).$$

Noting that the total number of observed customers equals 99, we can obtain the theoretical numbers of data falling into each class by $99 \times P\{a < X \leq b\}$. For instance, the theoretical number of data falling into the first class is calculated by $99 \times P\{0 < X \leq 10\} = 24.01$, where $1/\lambda = 36.20$, which has been calculated by the sample mean of the arrival times of the customers.

(iii) Gamma Distribution $X \sim GAM(\lambda, k)$

The density of the *gamma distribution* is given by

$$f_X(x) = \frac{\lambda^k x^{k-1} e^{-\lambda x}}{\Gamma(k)} \qquad (x \geq 0), \tag{2.4.12}$$

and is denoted by $X \sim GAM(\lambda, k)$, where λ $(\lambda > 0)$ and k $(k > 0)$ are parameters, and $\Gamma(k)$ is a gamma function of order k:

$$\Gamma(k) = \int_0^\infty e^{-x} x^{k-1} dx. \tag{2.4.13}$$

Integrating by parts, we have

$$\Gamma(k) = -e^{-x} x^{k-1} \Big|_0^\infty + (k-1) \int_0^\infty e^{-x} x^{k-2} dx = (k-1)\Gamma(k-1), \tag{2.4.14}$$

and

$$\Gamma(1) = \int_0^\infty e^{-x} dx = 1. \tag{2.4.15}$$

Thus, if k is a positive integer, $\Gamma(k) = (k-1)!$ Figure 2.4.3 shows the density of the gamma distribution $X \sim GAM(\lambda, k)$, where $\lambda = 1$; $k = 0.5$, 1, 2, 5.

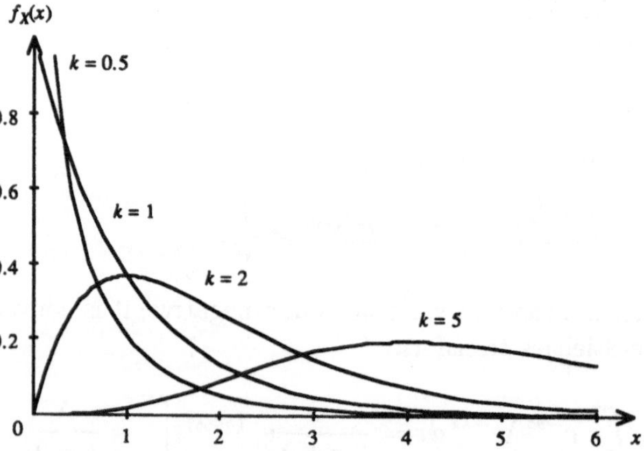

Fig. 2.4.3 The densities of $X \sim GAM(\lambda, k)$, where $\lambda = 1; k = 0.5, 1, 2, 5$.

The mean, variance, and characteristic function are given by

$$E[X] = \int_0^\infty x \frac{\lambda^k x^{k-1} e^{-\lambda x}}{\Gamma(k)} dx = \frac{k}{\lambda} \int_0^\infty \frac{\lambda^{k+1} x^k e^{-\lambda x}}{\Gamma(k+1)} dx = \frac{k}{\lambda}, \tag{2.4.16}$$

$$Var(X) = E[X^2] - E[X]^2 = \frac{k}{\lambda^2}, \tag{2.4.17}$$

and

$$\varphi_X(u) = \int_0^\infty e^{iux} \frac{\lambda^k x^{k-1} e^{-\lambda x}}{\Gamma(k)} dx = \left(\frac{\lambda}{\lambda - iu}\right)^k, \tag{2.4.18}$$

respectively. Since the gamma random variable is also non-negative, we can calculate the Laplace-Stieltjes transform:

$$F_X^*(s) = \int_0^\infty e^{-sx} \frac{\lambda^k x^{k-1} e^{-\lambda x}}{\Gamma(k)} dx = \left(\frac{\lambda}{s + \lambda}\right)^k. \tag{2.4.19}$$

It is finally noted that $GAM(\lambda, 1) = EXP(\lambda)$ which will be interpreted in the following section.

(iv) Weibull Distribution $X \sim WEI(\alpha, \beta)$

The density of the *Weibull distribution* is given by

$$f_X(x) = \alpha\beta(\alpha x)^{\beta-1} e^{-(\alpha x)^\beta} \qquad (x \geq 0), \tag{2.4.20}$$

and is denoted as $X \sim WEI(\alpha, \beta)$, where α $(\alpha > 0)$ and β $(\beta > 0)$ are parameters. The corresponding distribution is given by

$$F_X(x) = 1 - e^{-(\alpha x)^\beta}. \tag{2.4.21}$$

The Weibull distribution is well-known in reliability theory, where β and α are called the *shape* and *scale* parameters, respectively. Figure 2.4.4 shows the densities of the Weibull distribution $X \sim WEI(\alpha, \beta)$, where $\alpha = 1$; $\beta = 0.5, 1, 2, 5$.

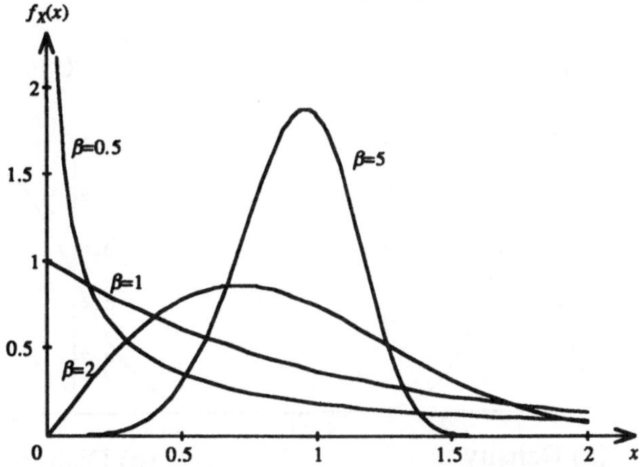

Fig. 2.4.4 The densities of $X \sim WEI(\alpha, \beta)$, where $\alpha = 1; \beta = 0.5, 1, 2, 5$.

The mean and variance can be expressed in terms of the gamma function defined in Eq.(2.4.13):

$$E[X] = \int_0^\infty x\alpha\beta(\alpha x)^{\beta-1}e^{-(\alpha x)^\beta}dx = \frac{1}{\alpha}\Gamma\left(1+\frac{1}{\beta}\right), \tag{2.4.22}$$

$$Var(X) = E[X^2] - E[X]^2 = \left(\frac{1}{\alpha}\right)^2\left\{\Gamma\left(1+\frac{2}{\beta}\right) - \Gamma\left(1+\frac{1}{\beta}\right)^2\right\}. \tag{2.4.23}$$

However, it is impossible to obtain the analytic forms of the characteristic function and Laplace-Stieltjes transform in general.

In particular, it is true that the Weibull distribution $WEI(\alpha, 2)$ (i.e., $\beta = 2$) is somtimes referred to as the *Rayleigh distribution.*

(v) Normal Distribution $X \sim N(\mu, \sigma^2)$

Normal distribution is well-known and plays a central role in statistics. As will be shown in the following section, the so-called "Central Limit Theorem" asserts that the sample mean tends toward normal distribution as sample size tends toward infinity.

The density of *normal distribution* is given by

$$f_X(x) = \frac{1}{\sqrt{2\pi}\sigma}\exp\left\{-\frac{(x-\mu)^2}{2\sigma^2}\right\} \qquad (-\infty < x < \infty), \tag{2.4.24}$$

where

$$E[X] = \int_{-\infty}^{\infty} xf_X(x)dx = \mu, \tag{2.4.25}$$

$$Var(X) = E[X^2] - E[X]^2 = \sigma^2. \tag{2.4.26}$$

Normal distribution with its density in Eq.(2.4.24) is denoted by $X \sim N(\mu, \sigma^2)$, where the mean μ and variance σ^2 (or standard deviation $\sigma > 0$) are parameters. Figure 2.4.5 shows the density and distribution of normal distribution $X \sim N(0, 1)$.

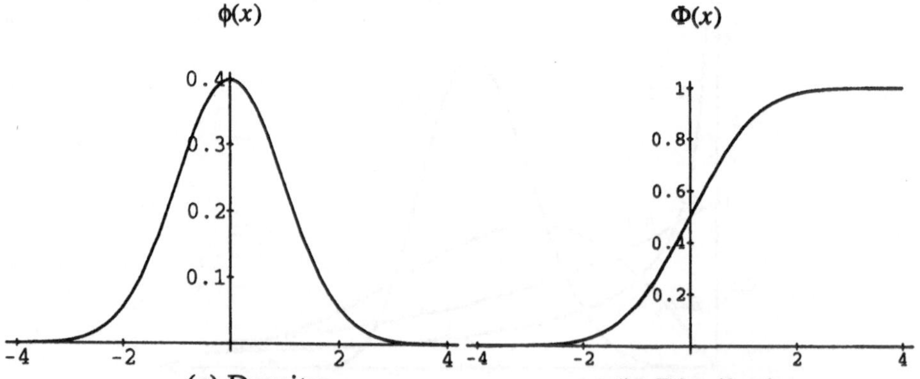

(a) Density (b) Distribution

Fig. 2.4.5 The density and distribution of $X \sim N(0,1)$.

The characteristic function is given by

$$\varphi_X(u) = \exp\left(i\mu u - \frac{\sigma^2 u^2}{2}\right) \tag{2.4.27}$$

As shown in Eq.(2.2.13), we should introduce the standardized random variable Y corresponding to X for normal distribution:

$$Y = \frac{X - E[X]}{\sqrt{Var(X)}} = \frac{X - \mu}{\sigma}, \tag{2.4.28}$$

with $E[Y] = 0$ and $Var(Y) = 1$. Then

$$\phi(y) = \frac{1}{\sqrt{2\pi}} e^{-\frac{y^2}{2}}, \tag{2.4.29}$$

where

$$\Phi(y) = \int_{-\infty}^{y} \phi(x) dx, \tag{2.4.30}$$

which is the distribution of standardized normal distribution $Y \sim N(0, 1)$. The numerical tables for calculating the cumulative probability in Eq.(2.4.30) are readily available.

(vi) Lognormal Distribution $X \sim LOGN(\mu, \sigma^2)$

The domain of normal distribution is $(-\infty, \infty)$. If $\log X$ $(X \geq 0)$ is distributed normally, the random variable X is distributed with the following density:

$$f_X(x) = \frac{1}{\sqrt{2\pi}\sigma x} \exp\left\{-\frac{(\log x - \mu)^2}{2\sigma^2}\right\} \qquad (x > 0), \tag{2.4.31}$$

which is the density of *lognormal distribution* and is denoted by $X \sim LOGN$ (μ, σ^2), where μ and σ $(\sigma > 0)$ are the parameters. The mean and variance are given by

$$E[X] = \exp\left(\mu + \frac{\sigma^2}{2}\right), \tag{2.4.32}$$

$$Var(X) = \exp\left(2\mu + 2\sigma^2\right) - \exp\left(2\mu + \sigma^2\right). \tag{2.4.33}$$

(vii) Beta Distribution $X \sim BETA(\alpha, \beta)$

The density of the *beta distribution* is given by

$$f_X(x) = \begin{cases} \frac{\Gamma(\alpha+\beta)}{\Gamma(\alpha)\Gamma(\beta)} x^{\alpha-1}(1-x)^{\beta-1} & (0 < x < 1) \\ 0 & (\text{elsewhere}), \end{cases} \tag{2.4.34}$$

and is denoted by $X \sim BETA(\alpha, \beta)$, where α ($\alpha > 0$) and β ($\beta > 0$) are the parameters and

$$B(\alpha, \beta) = \int_0^1 x^{\alpha-1}(1-x)^{\beta-1}dx \tag{2.4.35}$$

is the beta function of orders α and β. Note that

$$B(\alpha, \beta) = \frac{\Gamma(\alpha)\Gamma(\beta)}{\Gamma(\alpha+\beta)} \tag{2.4.36}$$

is the well-known relationship between the beta function in Eq.(2.4.35) and the gamma function in Eq.(2.4.13). Figure 2.4.6 shows the densities of the beta distribution $X \sim BETA(\frac{2}{3}, \frac{1}{3})$, $BETA(\frac{1}{2}, 1)$, $BETA(1, 2)$, $BETA(4, 2)$.

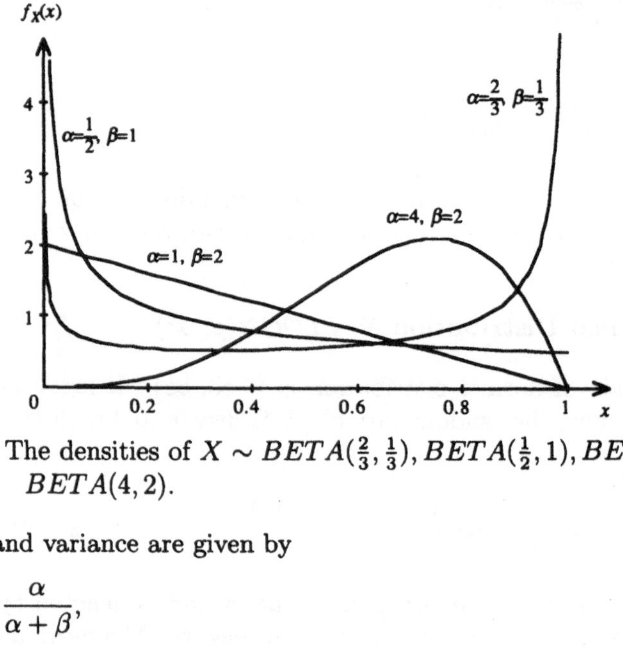

Fig. 2.4.6 The densities of $X \sim BETA(\frac{2}{3}, \frac{1}{3}), BETA(\frac{1}{2}, 1), BETA(1,2), BETA(4,2)$.

The mean and variance are given by

$$E[X] = \frac{\alpha}{\alpha+\beta}, \tag{2.4.37}$$

$$Var(X) = \frac{\alpha\beta}{(\alpha+\beta)^2(\alpha+\beta+1)}. \tag{2.4.38}$$

2.5 Multivariate Distributions

The random variables introduced in the preceding section were all functions that assign a real value (i.e., a one-dimensional real space) to each event. If a two-dimensional real space can be considered, the corresponding (probability) distribution is defined by

$$F_{XY}(x, y) = P\{X \leq x, Y \leq y\}. \tag{2.5.1}$$

Such a distribution is called the *bivariate joint distribution* of two random variables X and Y. The *marginal distribution* of the random variable X is defined by

$$\begin{aligned} F_X(x) &= P\{X \leq x\} \\ &= \lim_{y\to\infty} P\{X \leq x, Y \leq y\} \\ &= F_{XY}(x, \infty). \end{aligned} \tag{2.5.2}$$

The marginal distribution of Y is similarly defined by $F_Y(y) = F_{XY}(\infty, y)$.

The random variables X and Y are said to be *independent* if

$$F_{XY}(x, y) = F_X(x)F_Y(y) \tag{2.5.3}$$

for all x and y.

If the two random variables are discrete, the *joint probability mass function* of the two random variables is defined by

$$p_{XY}(x, y) = P\{X = x, Y = y\}, \tag{2.5.4}$$

and the corresponding joint distribution is given by

$$F_{XY}(x, y) = \sum_{a\leq x}\sum_{b\leq y} p_{XY}(a, b). \tag{2.5.5}$$

In addition, if

$$p_{XY}(x, y) = p_X(x)p_Y(y) \tag{2.5.6}$$

for all x and y, the two discrete random variables X and Y are *independent*, where

$$p_X(x) = \sum_{y=0}^{\infty} p_{XY}(x, y) \tag{2.5.7}$$

is the *marginal probability mass function* of X, and $p_Y(y)$ is the marginal probability mass function of Y, which is similarly defined.

If the two random variables are continuous, the joint distribution is given by

$$F_{XY}(x, y) = \int_{-\infty}^{x}\int_{-\infty}^{y} f_{XY}(\xi, \eta)\, d\xi d\eta, \tag{2.5.8}$$

where $f_{XY}(x, y)$ is called the *joint density* of X and Y. It is clear that the joint density

$$f_{XY}(x, y) = \frac{\partial^2 F(x, y)}{\partial x \partial y} \tag{2.5.9}$$

can be derived by partial differentiations of x and y. In addition, if

$$f_{XY}(x, y) = f_X(x)f_Y(y) \tag{2.5.10}$$

for all x and y, then two continuous random variables X and Y are independent, where

$$f_X(x) = \int_{-\infty}^{\infty} f_{XY}(x, y)dy \tag{2.5.11}$$

is the *marginal density* of X, and $f_Y(y)$ is the marginal density of Y, which is similarly defined.

The conditional probability $P\{A \mid B\}$ of the event A given the event B was defined by Eq.(1.3.13). Let X and Y be the discrete random variables whose joint probability mass functions $p_{XY}(x, y)$, and whose respective marginal probability mass functions $p_X(x)$ and $p_Y(y)$ are given. Then the *conditional probability mass function* $p_{X|Y}(x \mid y)$ of X given $Y = y$ is defined by

$$p_{X|Y}(x \mid y) = \frac{p_{XY}(x, y)}{p_Y(y)}, \tag{2.5.12}$$

provided $p_Y(y) > 0$. The conditional distribution of X given $Y = y$ is defined by

$$F_{X|Y}(x \mid y) = \sum_{a \le x} \frac{p_{XY}(a, y)}{p_Y(y)}, \tag{2.5.13}$$

provided $p_Y(y) > 0$. The *conditional expectation* of X given $Y = y$ is defined by

$$E[X \mid Y = y] = \sum_{x=0}^{\infty} \frac{x p_{XY}(x, y)}{p_Y(y)}, \tag{2.5.14}$$

provided $p_Y(y) > 0$. Applying the law of total probability, we have

$$p_X(x) = \sum_{y=0}^{\infty} p_{X|Y}(x \mid y)\, p_Y(y). \tag{2.5.15}$$

Taking the expectation of X in the above, we have

$$\begin{aligned} E[X] &= \sum_x x p_X(x) \\ &= \sum_{x=0}^{\infty} \sum_{y=0}^{\infty} x p_{X|Y}(x \mid y) p_Y(y) \\ &= E[E[X \mid Y]]. \end{aligned} \tag{2.5.16}$$

Let X and Y be the continuous random variables whose joint density $f_{XY}(x, y)$ in Eq.(2.5.9) and marginal densities $f_X(x)$ and $f_Y(y)$ are given. Then the *conditional density* $f_{X|Y}(x \mid y)$ of X given $Y = y$ is defined by

$$f_{X|Y}(x \mid y) = \frac{f_{XY}(x, y)}{f_Y(y)}, \tag{2.5.17}$$

provided $f_Y(y) > 0$. The *conditional distribution* of X given $Y = y$ is defined by

$$F_{X|Y}(x \mid y) = \int_{-\infty}^{x} \frac{f_{XY}(a, y)}{f_Y(y)} da. \tag{2.5.18}$$

The conditional expectation of X given $Y = y$ is defined by

$$E[X \mid Y = y] = \int_{-\infty}^{\infty} x \frac{f_{XY}(x, y)}{f_Y(y)} dx. \tag{2.5.19}$$

Applying the law of total probability, we have

$$f_X(x) = \int_{-\infty}^{\infty} f_{X|Y}(x \mid y) f_Y(y) dy. \tag{2.5.20}$$

Taking the expectation of X in Eq.(2.5.20), we have

$$\begin{aligned} E[X] &= \int_{-\infty}^{\infty} x f_X(x) dx \\ &= \int_{-\infty}^{\infty} \int_{-\infty}^{\infty} x f_{X|Y}(x \mid y) f_Y(y)\, dx dy \\ &= E[E[X \mid Y]], \end{aligned} \tag{2.5.21}$$

which was the same identity shown in Eq.(2.5.16) for the discrete random variables.

In the above discussions of the conditional probabilities, we have assumed that both of the two random variables are either discrete or continuous. However, we can easily generalize a case where one random variable is discrete and the other is continuous by introducing the appropriate probabilities. We omit the details here. If the two random variables X and Y are independent, the conditional probability mass function $p_{X|Y}(x \mid y)$ (or conditional density $f_{X|Y}(x \mid y)$) is independent of $Y = y$.

Consider the expectation or mean of $X + Y$. If the two random variables X and Y are discrete, then

$$\begin{aligned} E[X + Y] &= \sum_{x=0}^{\infty} \sum_{y=0}^{\infty} (x + y) p_{XY}(x, y) \\ &= \sum_{x=0}^{\infty} x p_X(x) + \sum_{y=0}^{\infty} y p_Y(y) \\ &= E[X] + E[Y]. \end{aligned} \tag{2.5.22}$$

The above identity can be similarly verified if the two random variables are continuous. In general, if k_0, k_1, and k_2 are constants, then

$$E[k_0 + k_1 X + k_2 Y] = k_0 + k_1 E[X] + k_2 E[Y]. \tag{2.5.23}$$

The variance of $X+Y$ is given by

$$\begin{aligned} Var(X+Y) &= E[((X+Y)-E[X+Y])^2] \\ &= E[(X-E[X]+Y-E[Y])^2] \\ &= E[(X-E[X])^2]+E[(Y-E[Y])^2]+2(E[XY]-E[X]E[Y]) \\ &= Var(X)+Var(Y)+2Cov(X,Y), \end{aligned} \tag{2.5.24}$$

where

$$\begin{aligned} Cov(X, Y) &= E[(X-E[X])(Y-E[Y])] \\ &= E[XY]-E[X]E[Y] \end{aligned} \tag{2.5.25}$$

is called the *covariance* of the two random variables X and Y. The *correlation coefficient* of X and Y is defined by

$$\rho(X, Y) = \frac{Cov(X, Y)}{\sqrt{Var(X)Var(Y)}}, \tag{2.5.26}$$

which is the dimensionless moment such that $-1 \leq \rho(X, Y) \leq 1$.

Example 2.5.1 (*Bivariate Normal Distribution*)

$$(X, Y) \sim N\left\{\begin{pmatrix} m_1 \\ m_2 \end{pmatrix}, \begin{pmatrix} \sigma_1^2 & \rho\sigma_1\sigma_2 \\ \rho\sigma_1\sigma_2 & \sigma_2^2 \end{pmatrix}\right\},$$

or

$$(X, Y) \sim N(m_1, m_2, \sigma_1^2, \sigma_2^2, \rho).$$

The density of bivariate normal distribution is given by

$$\begin{aligned} f_{XY}(x, y) &= \frac{1}{2\pi\sigma_1\sigma_2\sqrt{1-\rho^2}} \\ &\cdot \exp\left\{-\frac{1}{2(1-\rho^2)}\left[\left(\frac{x-m_1}{\sigma_1}\right)^2 - 2\rho\left(\frac{x-m_1}{\sigma_1}\right)\left(\frac{y-m_2}{\sigma_2}\right)+\left(\frac{y-m_2}{\sigma_2}\right)^2\right]\right\} \\ &\qquad (-\infty < x < \infty,\ -\infty < y < \infty), \end{aligned}$$

where

$$\begin{aligned} E[X] &= m_1, & E[Y] &= m_2, \\ Var(X) &= \sigma_1^2, & Var(Y) &= \sigma_2^2, \\ Cov(X, Y) &= \rho\sigma_1\sigma_2, & \rho(X, Y) &= \rho. \end{aligned}$$

The characteristic function is given by

$$E[e^{i(tX+uY)}] = \exp\left[im_1t + im_2u - \frac{1}{2}\left(\sigma_1^2t^2 + 2\rho\sigma_1\sigma_2tu + \sigma_2^2u^2\right)\right].$$

If $\rho = 0$, then X and Y are independent and

$$f_{XY}(x, y) = f_X(x) f_Y(y)$$
$$= \frac{1}{\sqrt{2\pi}\sigma_1} \exp\left[-\frac{1}{2}\left(\frac{x - m_1}{\sigma_1}\right)^2\right] \cdot \frac{1}{\sqrt{2\pi}\sigma_2} \exp\left[-\frac{1}{2}\left(\frac{y - m_2}{\sigma_2}\right)^2\right].$$

However, in general, it is not true that $\rho = 0$ implies that X and Y are independent. Figure 2.5.1 shows the density of bivariate normal distribution, when $\rho = 1/2,\ m_1 = m_2 = 0,\ \sigma_1^2 = \sigma_2^2 = 1$.

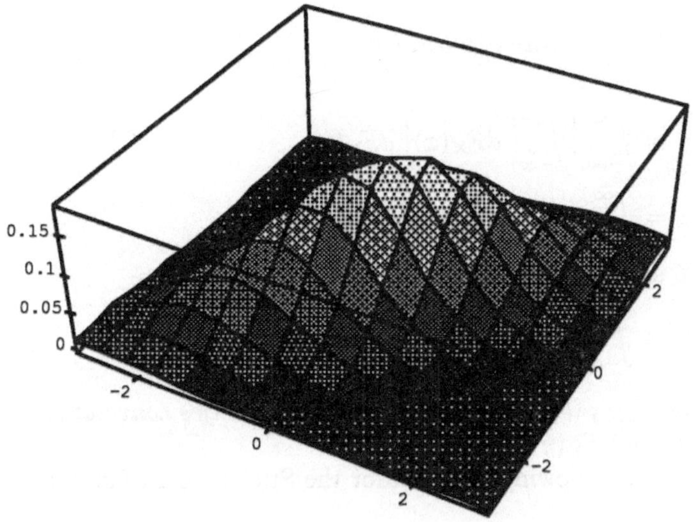

Fig. 2.5.1 The density of bivariate normal distribution, when $\rho = \frac{1}{2}, m_1 = m_2 = 0, \sigma_1^2 = \sigma_2^2 = 1$.

Example 2.5.2 (*Bivariate Exponential Distribution*) The bivariate exponential distribution of the two non-negative random variables X and Y is given by its survival probability

$$\bar{F}_{XY}(x, y) = P\{X > x, Y > y\}$$
$$= \exp[-\lambda_1 x - \lambda_2 y - \lambda_{12} \max(x, y)],$$

where $x \geq 0,\ y \geq 0,\ \lambda_1 \geq 0,\ \lambda_2 \geq 0$, and $\lambda_{12} \geq 0$. The marginal distributions of X and Y are

$$F_X(x) = 1 - \exp[-(\lambda_1 + \lambda_{12})x],$$
$$F_Y(y) = 1 - \exp[-(\lambda_2 + \lambda_{12})y],$$

both of which are exponential distributions. If $\lambda_{12} = 0$, then

$$F_{XY}(x, y) = F_X(x) F_Y(y)$$

for all x and y, which implies that X and Y are independent. On the other hand, if $\lambda_{12} > 0$, then X and Y are dependent and

$$\rho(X, Y) = \frac{\lambda_{12}}{\lambda_1 + \lambda_2 + \lambda_{12}}.$$

Let X and Y be *independent* random variables with distributions $F_X(x)$ and $F_Y(y)$. Then, the distribution of $X+Y$

$$\begin{aligned} F_{X+Y}(\xi) &= P\{X+Y \leq \xi\} \\ &= \int_{x+y\leq\xi}\int dF_X(x)dF_Y(y) \\ &= \int_{-\infty}^{\infty}\left[\int_{-\infty}^{\xi-y} dF_X(x)\right] dF_Y(y) \\ &= \int_{-\infty}^{\infty} F_X(\xi - y)\, dF_Y(y). \end{aligned} \tag{2.5.27}$$

By reversing X and Y, we also have

$$F_{X+Y}(\xi) = \int_{-\infty}^{\infty} F_Y(\xi - x)\, dF_X(x). \tag{2.5.28}$$

The distribution $F_{X+Y}(\xi)$ of $X+Y$ is called the *Stieltjes convolution* (or *convolution*) of $F_X(x)$ and $F_Y(y)$.

We often use the following notation for the Stieltjes convolution:

$$F_{X+Y}(\xi) = F_X * F_Y(\xi) = F_Y * F_X(\xi). \tag{2.5.29}$$

Let us consider a simple case. Let $X_1, X_2, \cdots, X_n$ be independent and identically distributed non-negative random variables with identical distribution $F(t)$ $(t \geq 0)$. Let $S_n = X_1 + X_2 + \cdots + X_n$ be the sum of the random variables $X_1, X_2, \cdots, X_n$. Then we have

$$P\{S_2 \leq t\} = \int_0^t F(t-x)\, dF(x) = F * F(t) = F^{(2)}(t), \tag{2.5.30}$$

$$P\{S_3 \leq t\} = \int_0^t F^{(2)}(t-x)\, dF(x) = F^{(2)} * F(t) = F^{(3)}(t), \tag{2.5.31}$$

and, in general,

$$P\{S_n \leq t\} = \int_0^t F^{(n-1)}(t-x)\, dF(x) = F^{(n-1)} * F(t) = F^{(n)}(t)$$

$$(n = 1, 2, \cdots), \tag{2.5.32}$$

where $F^{(0)}(t) = 1(t)$ (a step function) and $F^{(1)}(t) = F(t)$. $F^{(n)}(t)$ is called the *n-fold convolution* of itself and will be frequently used in renewal theory in Chapter 4.

Let $\phi_X(u)$, $\phi_Y(u)$, and $\phi_{X+Y}(u)$ be the characteristic functions of the random variables X, Y, and $X+Y$, respectively, where we assume that X and Y are independent. If X and Y are continuous random variables, then

$$\begin{aligned}
\phi_{X+Y}(u) &= \int_{-\infty}^{\infty} e^{iu\xi} \frac{d}{d\xi} \left[\int_{-\infty}^{\infty} F_X(\xi - y)\, dF_Y(y) \right] d\xi \\
&= \int_{-\infty}^{\infty} \int_{-\infty}^{\infty} e^{iu(\xi-y)} e^{iuy} \frac{d}{d\xi} F_X(\xi - y)\, d\xi dF_Y(y) \\
&= \int_{-\infty}^{\infty} e^{iu(\xi-y)} \frac{d}{d\xi} F_X(\xi - y) d\xi \int_{-\infty}^{\infty} e^{iuy}\, dF_Y(y) \\
&= \int_{-\infty}^{\infty} e^{iuz}\, dF_X(z) \int_{-\infty}^{\infty} e^{iuy}\, dF_Y(y) \qquad (\xi - y = z) \\
&= \phi_X(u)\phi_Y(u). \qquad (2.5.33)
\end{aligned}$$

That is, if X and Y are independent, the characteristic function of $X+Y$ is the product of the characteristic functions of X and Y. Noting Theorem 2.2.2, we obtain the distribution of $X+Y$ from the characteristic function $\phi_{X+Y}(u)$. In particular, if we are concerned with the non-negative random variables, we will use the Laplace-Stieltjes transforms instead of the characteristic functions.

Example 2.5.3 Let X and Y be independent Poisson random variables with parameters λ and μ, respectively. Note that X and Y are discrete non-negative random variables. The probability mass function of $X+Y$ is given by

$$\begin{aligned}
p_{X+Y}(n) &= P\{X+Y=n\} \\
&= \sum_{k=0}^{n} P\{X=k\}P\{Y=n-k\} \\
&= \sum_{k=0}^{n} e^{-\lambda}\frac{\lambda^k}{k!} \cdot e^{-\mu}\frac{\mu^{n-k}}{(n-k)!} \\
&= e^{-(\lambda+\mu)} \frac{1}{n!} \sum_{k=0}^{n} \frac{n!}{k!(n-k)!} \lambda^k \mu^{n-k} \\
&= e^{-(\lambda+\mu)} \frac{(\lambda+\mu)^n}{n!} \qquad (n = 0, 1, , \cdots),
\end{aligned}$$

which is also the probability mass function of the Poisson distribution with parameter $\lambda + \mu$. We next apply the characteristic function.

$$\begin{aligned}
\phi_{X+Y}(u) &= \phi_X(u)\phi_Y(u) \\
&= \exp\left[\lambda\left(e^{iu}-1\right)\right] \exp\left[\mu\left(e^{iu}-1\right)\right] \\
&= \exp\left[(\lambda+\mu)\left(e^{iu}-1\right)\right],
\end{aligned}$$

which is the characteristic function of the Poisson distribution with parameter $\lambda + \mu$.

Example 2.5.4 Let $X_1, X_2, \cdots, X_k$ be independent and identically distributed exponential random variables with parameter λ. Let $S_k = X_1 + X_2 + \cdots + X_k$ be the sum of the random variables $X_1, X_2, \cdots, X_k$. The characteristic function of S_k is given by

$$\begin{aligned}\phi_{S_k}(u) &= \phi_{X_1}(u)\phi_{X_2}(u)\cdots\phi_{X_k}(u)\\ &= \left(\frac{\lambda}{\lambda - iu}\right)^k,\end{aligned}$$

which is the characteristic function of the gamma distribution $S_k \sim GAM(\lambda, k)$.

Example 2.5.5 Let $X_1, X_2, \cdots, X_n$ be independent and identically distributed Bernoulli random variables with parameter p. Let $S_n = X_1 + X_2 + \cdots + X_n$ be the sum of the random variables $X_1, X_2, \cdots, X_n$. Note that the characteristic function of the random variable X_k is

$$\phi_{X_k}(u) = pe^{iu\cdot 1} + (1-p)e^{iu\cdot 0} = pe^{iu} + q \qquad (k = 1, 2, \cdots, n).$$

Then, the characteristic function of the random variable S_n is

$$\phi_{S_n}(u) = \prod_{k=1}^{n} \phi_{X_k}(u) = [\phi_{X_1}(u)]^n = \left(pe^{iu} + q\right)^n,$$

which is the characteristic function of the binomial distribution $S_n \sim B(n, p)$.

2.6 Limit Theorems

In this section we show a few limit theorems which will be used later. We first show the following inequality.

Theorem 2.6.1 (*Markov's Inequality*) Let X be a non-negative random variable. For any $t > 0$, we have

$$P\{X \geq t\} \leq \frac{E[X]}{t}. \tag{2.6.1}$$

Proof

$$\begin{aligned}E[X] &= \int_0^\infty x\, dF_X(x)\\ &= \int_0^t x\, dF_X(x) + \int_t^\infty x\, dF_X(x)\\ &\geq \int_t^\infty x\, dF_X(x)\\ &\geq t\int_t^\infty dF_X(x) = tP\{X \geq t\}. \end{aligned} \tag{2.6.2}$$

Applying Markov's inequality in Eq.(2.6.1), the following well-known inequality results:

Theorem 2.6.2 (*Chebyshev's Inequality*) Let X be the random variable with finite mean μ and variance σ^2. For any $k > 0$,

$$P\{|X-\mu| \geq k\} \leq \frac{\sigma^2}{k^2}. \tag{2.6.3}$$

Proof Noting that $(X-\mu)^2$ is a non-negative random variable, we have

$$P\{(X-\mu)^2 \geq k^2\} \leq \frac{E[(X-\mu)^2]}{k^2} = \frac{\sigma^2}{k^2}, \tag{2.6.4}$$

which implies Eq.(2.6.3).

In particular, if we assume $k = m\sigma$ (m times the standard deviation, where $m > 1$), we can rewrite the alternative Chebyshev's inequality

$$P\{|X-\mu| \geq m\sigma\} \leq \frac{1}{m^2}. \tag{2.6.5}$$

Example 2.6.1 In Eq.(2.6.5), if we assume $m = 3$, then

$$P\{|X-\mu| \geq 3\sigma\} \leq \frac{1}{9} \doteq 0.111.$$

If we further assume that $X \sim N(\mu, \sigma^2)$, then

$$P\{|X-\mu| \geq 3\sigma\} \doteq 1 - 0.997 = 0.003.$$

However, Chebyshev's inequality asserts that the probability $P\{|X-\mu| \geq 3\sigma\}$ is less than or equal to 0.111 irrespective of the distributions.

We derive two limit theorems which are important to the latter discussions. However, we omit the proofs for these two limit theorems.

Theorem 2.6.3 (*Strong Law of Large Numbers*) Let $X_1, X_2, \cdots$ be independent and identically distributed random variables with mean $E[X_i] = \mu$ ($i = 1, 2, \cdots$). With probability 1, as $n \to \infty$,

$$\frac{X_1 + X_2 + \cdots + X_n}{n} \longrightarrow \mu. \tag{2.6.6}$$

This means that

$$P\left\{\lim_{n\to\infty} \frac{X_1 + X_2 + \cdots + X_n}{n} = \mu\right\} = 1, \tag{2.6.7}$$

or, equivalently, for every $\varepsilon > 0$ and $\delta > 0$, there exists an integer $N = N(\varepsilon, \delta)$ such that

$$P\left\{\left|\frac{X_1 + X_2 + \cdots + X_n}{n} - \mu\right| < \varepsilon \text{ for every } n \geq N\right\} \geq 1 - \delta. \tag{2.6.8}$$

Theorem 2.6.3 is well-known and is called the *Strong Law of Large Numbers.* There are several variations of such a theorem. Theorem 2.6.3 is one of the simplest ones.

We now present the following theorem.

Theorem 2.6.4 (*Central Limit Theorem*) Let $X_1, X_2, \cdots$ be independent and identically distributed random variables with mean $E[X_i] = \mu$ $(i = 1, 2, \cdots)$ and variance $Var(X_i) = \sigma^2$. Then the *sample mean* is defined by

$$Y_n = \frac{X_1 + X_2 + \cdots + X_n}{n}. \tag{2.6.9}$$

The *standardized* random variable is defined by

$$Z_n = \frac{Y_n - \mu}{\frac{\sigma}{\sqrt{n}}} \tag{2.6.10}$$

with $E[Z_n] = 0$ and $Var(Z_n) = 1$. Then, as $n \to \infty$,

$$P\{Z_n \leq k\} \longrightarrow \Phi(k) = \frac{1}{\sqrt{2\pi}} \int_{-\infty}^{k} e^{-\frac{x^2}{2}} \, dx. \tag{2.6.11}$$

The central limit theorem asserts that the sample mean $Y_n = (X_1 + X_2 + \cdots + X_n)/n$ tends toward the normal distribution $Y_n \sim N(\mu, \frac{\sigma^2}{n})$ as $n \to \infty$, or equivalently, the standardized sample mean Z_n tends toward the standardized normal distribution $Z_n \sim N(0, 1)$ as $n \to \infty$.

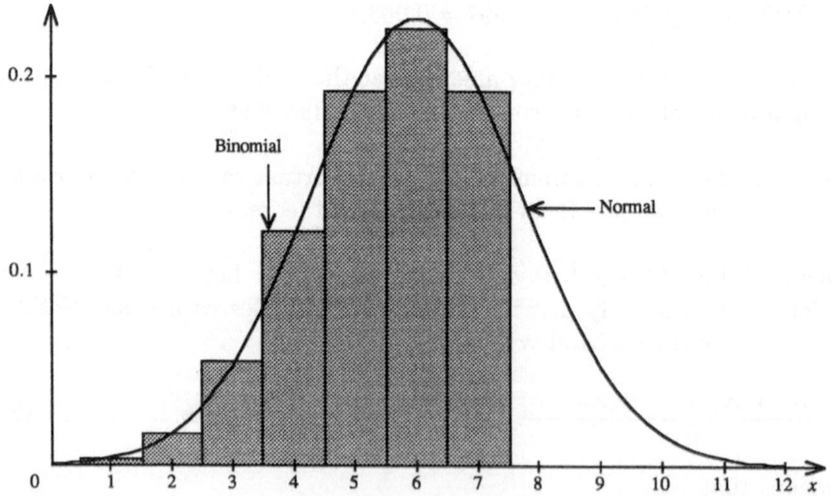

Fig. 2.6.1 The normal approximation for the binomial distribution $S_{12} \sim B(12, \frac{1}{2})$.

Example 2.6.2 As shown in Example 2.5.5, we derived that the random variable $S_n \sim B(n, p)$. Let us consider the fair coin tossing. The probability that heads appear less than or equal to 7 times out of 12 trials is given by

$$P\{0 \le X \le 7\} = \sum_{k=0}^{7} \binom{12}{k} \left(\frac{1}{2}\right)^k \left(\frac{1}{2}\right)^{12-k} = \frac{3302}{4096} = 0.8062.$$

On the other hand, we apply the central limit thereorem and calculate the corresponding probability by approximation. Noting that $E[S_{12}] = 12 \cdot \frac{1}{2} = 6$, $Var(S_{12}) = 12 \cdot \frac{1}{2} \cdot \frac{1}{2} = 3$, we can derive the corresponding probability by the following approximation (see Fig. 2.6.1):

$$P\{-0.5 < X \le 7.5\}.$$

Using the standardized random variable $Y = (X - 6)/\sqrt{3}$, we have

$$\begin{aligned} &P\left\{\frac{-0.5-6}{\sqrt{3}} < Y \le \frac{7.5-6}{\sqrt{3}}\right\} \\ &= P\{-3.75 < Y \le 0.87\} \\ &= \Phi(0.87) - \Phi(-3.75) \\ &= 0.8078 \end{aligned}$$

which is quite a good approximation for $P\{0 \le X \le 7\} = 0.8062$.

2.7 Problems 2

2.1 Verify the following sums of powers of integers:

(i) $1+2+\cdots+n=\frac{n(n+1)}{2}$.

(ii) $1^2+2^2+\cdots+n^2=\frac{n(n+1)(2n+1)}{6}$.

(iii) $1^3+2^3+\cdots+n^3=\left[\frac{n(n+1)}{2}\right]^2$.

2.2 Let X be a non-negative random variable with distribution $F_X(x)$. Assuming that $F_X(0)=0$, verify the following formulas:

$$E[X^n]=\int_0^\infty x^n dF_X(x)=n\int_0^\infty x^{n-1}[1-F_X(x)]dx \qquad (n=1,\,2,\,\cdots).$$

In particular, we have

$$E[X]=\int_0^\infty [1-F_X(x)]dx,$$

$$E[X^2]=2\int_0^\infty x[1-F_X(x)]dx.$$

2.3 The characteristic function $\varphi_X(u)$ of the uniform distribution $X\sim U(C+L,\,C+NL)$ is given in Eq.(2.3.5). Calculate the mean and variance of the uniform distribution by applying Eq.(2.2.18).

2.4 (*Example 1.4.5*) Show the following: For the hypergeometric distribution, if n is large and $n_1/n=p$, then the probability P_k in Example 1.4.5 is bounded by the following binomial distributions:

$$\binom{r}{k}\left(p-\frac{k}{n}\right)^k\left(q-\frac{r-k}{n}\right)^{r-k}<P_k<\binom{r}{k}p^kq^{r-k}\left(1-\frac{r}{n}\right)^{-r}.$$

That is, for large populations, the hypergeometric distribution is approximated by the binomial distribution with $p=n_1/n$.

2.5 Calculate the mean and variance by applying the characteristic function of the negative binomial distribution $X\sim NB(p,r)$.

2.6 (i) Calculate the mean and variance of the exponential distribution $X\sim EXP(\lambda)$ by applying Problem 2.2 above.

(ii) Calculate the mean and variance of the exponential distribution $X\sim EXP(\lambda)$ by applying Eq.(2.2.18).

2.7 (*Example 2.4.1*) Calculate the theoretical numbers of data falling into each class by $99\times P\{10n<X\le 10(n+1)\}$ $(n=1,\,2,\,3,\,\cdots,\,11)$, and compare Table 2.1.2.

2.8 Calculate the mean, variance, and Laplace-Stieltjes transform of the gamma distribution $X \sim GAM(\lambda, k)$.

2.9 Calculate the mean and variance of the Weibull distribution $X \sim WEI(\alpha, \beta)$ (Hint: Use the gamma function in Eq.(2.4.13)).

2.10 For the normal distribution $X \sim N(0, 1)$, show that $\Phi(\infty) = 1$, i.e., $\phi(x)$ is the density of the standard normal distribution $X \sim N(0, 1)$.

2.11 Calculate the mean and variance of the lognormal distribution $X \sim LOGN$ (μ, σ^2).

2.12 Calculate the mean and variance of the beta distribution $X \sim BETA(\alpha, \beta)$.

2.13 (*Example 2.5.1*) The characteristic function of the bivariate normal distribution $\varphi_{XY}(t, u) = E[e^{i(tX+uY)}]$ is given in Example 2.5.1. Show the mean and variance of X and covariance by using the following formulas:

$$m_1 = \frac{1}{i}\frac{\partial \varphi_{XY}(t, u)}{\partial t}\bigg|_{t=0,\, u=0}.$$

$$E[X^2] = (-1)\,\frac{\partial^2 \varphi_{XY}(t, u)}{\partial t^2}\bigg|_{t=0,\, u=0}.$$

$$E[XY] = (-1)\,\frac{\partial^2 \varphi_{XY}(t, u)}{\partial t \partial u}\bigg|_{t=0,\, u=0}.$$

2.14 (*Example 2.5.2*) (i) Calculate the means and variances of X and Y for a bivariate exponential distribution by applying Problem 2.2. (ii) Calculate the covariance $Cov(X, Y)$ and $\rho(X, Y)$ by using the following formula:

$$E[XY] = \int_0^\infty \int_0^\infty \bar{F}_{XY}(x, y)\, dx dy.$$

2.15 Let $X_1, X_2, \cdots, X_n$ denote the independent and identically distributed normal random variables, i.e., $X_i \sim N(\mu, \sigma^2)$ $(i = 1, 2, \cdots, n)$. The sample mean is defined by

$$Y = \frac{X_1 + X_2 + \cdots + X_n}{n}.$$

Show that $Y \sim N(\mu, \frac{\sigma^2}{n})$.

2.16 Let X and Y be the continuous random variables with bivariate normal distribution:

$$(X, Y) \sim N\left\{\begin{pmatrix} m_1 \\ m_2 \end{pmatrix}, \begin{pmatrix} \sigma_1^2 & \rho\sigma_1\sigma_2 \\ \rho\sigma_1\sigma_2 & {\sigma_2}^2 \end{pmatrix}\right\},$$

(ref. Example 2.5.1). Show that the conditional distribution of Y given that $X = x$ is distributed normal with mean $m_1 + \rho\sigma_2(x - m_1)/\sigma_1$ and variance ${\sigma_2}^2(1 - \rho^2)$.

Chapter 3

Poisson Processes

3.1 Stochastic Processes

As shown in Section 2.1, a stochastic process can be described by the laws of probability at each point of time $t \geq 0$. As shown in Fig. 2.1.1, we are very much interested in the random variable $N(t)$, which denotes the number of arriving customers up to time t, where $N(t) = 0, 1, 2, \cdots$. A *counting process* $\{N(t), t \geq 0\}$ is one of the stochastic processes, and Fig. 2.1.1 shows a "sample function" or "sample path" of the counting process $\{N(t), t \geq 0\}$. We can consider several examples of counting processes, where the "customer" is replaced by other relevant words such as the "call" in congestion theory, the "failure" of machines, and the arriving "job" or arriving "transaction" of computer systems.

The Poisson process in this chapter and the renewal process in the following chapter are both counting processes $\{N(t), t \geq 0\}$. It is of our interest to discuss the behavior of the random variable $N(t)$ from several viewpoints.

The random phenomena can be described by the Poisson process, a process that is interesting and has simple properties from the viewpoint of mathematics. The Poisson process will be precisely defined in the following section. We will now just sketch the Poisson process informally. Let us consider the time interval $[0, t]$ which is divided by n equally small time intervals, where $n\Delta t = t$. For each small interval $[k\Delta t, (k+1)\Delta t]$ $(k = 0, 1, 2, \cdots, n-1)$, we consider the Bernoulli trials with probability p, i.e., the probability of a customer arriving is p, and the probability of a customer *not* arriving is $q = 1 - p$, where we assume that never do two or more customers arrive in any small interval. The probability that k

customers arrive in the time interval $[0, t]$ is given by

$$\binom{n}{k} p^k q^{n-k} = \frac{\left(1-\frac{1}{n}\right)\left(1-\frac{2}{n}\right)\cdots\left(1-\frac{k-1}{n}\right)}{k!}(\lambda t)^k \left(1-\frac{\lambda t}{n}\right)^{n-k}$$

$$\to P\{N(t)=k\} = \frac{(\lambda t)^k}{k!} e^{-\lambda t} \qquad (k = 0,\ 1,\ 2,\ \cdots), \tag{3.1.1}$$

as $n \to \infty$, where $p = \lambda \Delta t$ (ref. Example 2.3.3). That is, the probability $P\{N(t) = k\}$ obeys the Poisson distribution $N(t) \sim POI(\lambda t)$.

Under the same assumptions as above, we derive the probability that the first customer arrives at the nth small interval. Noting the geometric distribution, we have

$$pq^{n-1} = \lambda \Delta t \left(1-\frac{\lambda t}{n}\right)^{n-1} \to \lambda e^{-\lambda t} \Delta t, \tag{3.1.2}$$

as $n \to \infty$, where $p = \lambda \Delta t$ and $t = n \Delta t$. That is, the probability $P\{X_1 \leq t\}$ that the first customer arrives up to time t is given by

$$dP\{X_1 \leq t\} = \lambda e^{-\lambda t} dt, \tag{3.1.3}$$

or

$$P\{X_1 \leq t\} = 1 - e^{-\lambda t}, \tag{3.1.4}$$

which is the exponential distribution $X_1 \sim EXP(\lambda)$. Repeating similar discussions of the arriving time X_k for the kth customer $(k = 2,\ 3,\ \cdots)$ measured from the preceding customer's arriving time as the time origin, we have

$$P\{X_k \leq t\} = 1 - e^{-\lambda t} \qquad (k = 2,\ 3,\ \cdots). \tag{3.1.5}$$

That is, the random variables X_k $(k = 1,\ 2,\ \cdots)$ are independent and identically distributed with distributions $X_k \sim EXP(\lambda)$.

We describe the Poisson process from two different viewpoints: First, the random variable $N(t)$ obeys the Poisson distribution $N(t) \sim POI(\lambda t)$. Secondly, the random variables X_k $(k = 1,\ 2,\ \cdots)$ obey the common exponential distributions $X_k \sim EXP(\lambda)$, where all random variables are independent.

We generally discuss a stochastic process $\{X(t),\ t \in T\}$ which is a family of random variables, where $X(t)$ describes the law of probability at time $t \in T$. We restrict ourselves to the non-negative time parameter $t \geq 0$ for the index set T throughout this book.

If the time parameter T is a countable set $T = \{0,\ 1,\ 2,\ \cdots\}$, the process $\{X(n),\ n = 0,\ 1,\ 2,\ \cdots\}$ is called a *discrete-time stochastic process*, and if T is a continuum, the process $\{X(t),\ t \geq 0\}$ is called a *continuous-time stochastic process*.

For a stochastic process $\{X(t),\ t \in T\}$, a set of all possible values of $X(t)$ is called a *state space*. Throughout this book we restrict ourselves to a *discrete state space* which can be described by the non-negative integers $i = 0,\ 1,\ 2,\ \cdots$, unless otherwise specified. Of course, we are also interested in a stochastic process $\{X(t),\ t \in T\}$ with a *continuous state space*. An example of the latter is a Brownian motion or diffusion process; this represents a different interest. However, we do not intend to discuss such a stochastic process with a continuous state space in this book.

A continuous-time stochastic process $\{X(t),\ t \geq 0\}$ is said to have *independent increments* if, for all $0 \leq t_0 < t_1 < \cdots < t_n$, the variables

$$X(t_1) - X(t_0),\ X(t_2) - X(t_1),\ \cdots,\ X(t_n) - X(t_{n-1})$$

are independent, where the difference $X(t_i) - X(t_{i-1})$ is called the *increment*. The process $\{X(t),\ t \geq 0\}$ is said to have *stationary increments* if $X(t+s) - X(t)$ has the same distribution for all $t \geq 0$. The process $\{X(t),\ t \geq 0\}$ is said to have *stationary independent increments* if $X(t_2 + s) - X(t_1 + s)$ has the same distribution for all $t_2 > t_1 \geq 0$ and $s > 0$. A *sample function* or *sample path* is a realization of the stochastic process, an example of which was given in Fig. 2.1.1.

In this chapter we develop the (homogeneous) Poisson process which has stationary independent increments. We also discuss the nonhomogeneous Poisson process which has only independent increments.

3.2 The Poisson Process

A counting process $\{N(t),\ t \geq 0\}$ is one of the simplest stochastic processes and represents the number of *events*, where the events can refer to arriving customers, particles, jobs, transactions, data, and so on. Before introducing the definition of the Poisson process, we introduce the following:

Definition 3.2.1 The function $f(h)$ is said to be $o(h)$ if

$$\lim_{h \to 0} \frac{f(h)}{h} = 0. \tag{3.2.1}$$

Example 3.2.1 For a small (time) interval ($h > 0$), we have

$$1 - e^{-\lambda h} = \lambda h - \frac{(\lambda h)^2}{2!} + \frac{(\lambda h)^3}{3!} - \cdots = \lambda h + o(h),$$

$$e^{-\lambda h} = 1 - \lambda h + \frac{(\lambda h)^2}{2!} - \frac{(\lambda h)^3}{3!} + \cdots = 1 - \lambda h + o(h).$$

That is, the equations above show that the probability that an event takes place for a small interval $h > 0$ is $\lambda h + o(h)$ and that no event takes place for a small interval $h > 0$ is $1 - \lambda h + o(h)$ as $h \to 0$, respectively. Note that $P\{X \le t\} = 1 - e^{-\lambda t}$ and $P\{X > t\} = e^{-\lambda t}$ for the exponential distribution $X \sim EXP(\lambda)$.

We are ready to define the Poisson process as follows:

Definition 3.2.2 A counting process $\{N(t),\ t \ge 0\}$ is called a *Poisson process* with parameter $\lambda > 0$ if the following conditions are satisfied:

(i) $N(0) = 0$.

(ii) The process has stationary independent increments.

(iii) $P\{N(h) = 1\} = \lambda h + o(h)$.

(iv) $P\{N(h) \ge 2\} = o(h)$.

From Definition 3.2.2 for the Poisson process, we derive the following probability: For any $t \ge 0$,

$$P_k(t) = P\{N(t) = k|\ N(0) = 0\} \qquad (k = 0, 1, 2, \cdots), \tag{3.2.2}$$

which denotes the probability that k events take place for the time interval $(0, t]$. Noting that $P_k(t)$ is the probability mass function for a fixed t, we have for a fixed t

$$\sum_{k=0}^{\infty} P_k(t) = 1, \tag{3.2.3}$$

since the total probability is unity. Applying the stationarity of the Poisson process, we have

$$P\{N(s+t) - N(s) = k\} = P\{N(t) = k|\ N(0) = 0\} = P_k(t), \tag{3.2.4}$$

for any $t \ge 0$ and $s \ge 0$. Applying (ii), (iii), and (iv) in Definition 3.2.2, we have

$$P\{N(t+h) - N(t) = k|N(t) = i\}$$
$$= \begin{cases} 1 - \lambda h + o(h) & (k = i) \\ \lambda h + o(h) & (k = i+1) \\ o(h) & (k > i+1). \end{cases} \tag{3.2.5}$$

Using Eq.(3.2.5) and Definition 3.2.2, we have

$$\begin{aligned} P_0(t+h) &= P_0(t)P_0(h) \\ &= P_0(t)[1 - \lambda h + o(h)] \end{aligned} \tag{3.2.6}$$

and

$$P_k(t+h) = P_{k-1}(t)[\lambda h + o(h)] + P_k(t)[1 - \lambda h + o(h)] + \sum_{i=2}^{k} P_{k-i}(t)o(h),$$

$$(k = 1, 2, 3, \cdots). \quad (3.2.7)$$

Rearranging Eqs.(3.2.6) and (3.2.7) and assuming $h \to 0$ imply the following set of simultaneous differential equations:

$$\frac{dP_0(t)}{dt} = -\lambda P_0(t), \tag{3.2.8}$$

$$\frac{dP_k(t)}{dt} = \lambda P_{k-1}(t) - \lambda P_k(t), \qquad (k = 1, 2, \cdots), \tag{3.2.9}$$

with the initial conditions $P_0(0) = P\{N(0) = 0\} = 1$ and $P_k(0) = P\{N(0) = k\} = 0$ $(k = 1, 2, \cdots)$, which can be derived from (i) of Definition 3.2.2. Applying the theory of differential equations, we have

$$P_k(t) = \frac{(\lambda t)^k}{k!} e^{-\lambda t} \qquad (k = 0, 1, 2, \cdots), \tag{3.2.10}$$

which is the probability mass function of the Poisson distribution with parameter λt, i.e., for a fixed t, $N(t) \sim POI(\lambda t)$. The rigorous verification of deriving Eq.(3.2.10) requires the techniques of mathematical induction or generating functions (see Problems 3.1 and 3.2).

We show an alternative definition of the Poisson process.

Definition 3.2.3 A counting process $\{N(t),\ t \geq 0\}$ is called a *Poisson process* with parameter $\lambda > 0$ if the following conditions are satisfied:

(i) $N(0) = 0$.

(ii) The process has independent increments.

(iii) The probability that k events take place for any interval t is given by

$$P\{N(t+s) - N(s) = k\} = \frac{(\lambda t)^k}{k!} e^{-\lambda t} \qquad (k = 0, 1, 2, \cdots),$$

for all $s,\ t \geq 0$. That is, $N(t+s) - N(s) \sim POI(\lambda t)$ for all $s,\ t \geq 0$.

As just shown, we have verified that Definition 3.2.2 implies Definition 3.2.3. We will verify that Definition 3.2.3 implies Definition 3.2.2. It is obvious that the process has stationary increments from (iii) of Definition 3.2.3. We further show that

$$P\{N(h) = 1\} = \lambda h e^{-\lambda h} = \lambda h + o(h), \tag{3.2.11}$$

$$P\{N(h) = k\} = \frac{(\lambda h)^k}{k!} e^{-\lambda h} = o(h) \qquad (k = 2, 3, \cdots), \tag{3.2.12}$$

from the fact that $N(h) \sim POI(\lambda h)$. From the facts above, we verify that Definition 3.2.3 implies Definition 3.2.2. That is, Definitions 3.2.2 and 3.2.3 are equivalent.

Example 3.2.2 The customers arriving at a shop obey a Poisson process with 2 persons per hour during business hours from 10 *a.m.* ($t = 0$) to 6 *p.m.* ($t = 8$). (1) Calculate the probabilities that k customers ($k = 0, 1, 2$) arrive from 1 *p.m.* to 3 *p.m.* Noting that the Poisson process has stationary increments and $\lambda = 2$ persons/hour, we have for $t = 2$ (from 1 *p.m.* to 3 *p.m.*)

$$P_0(2) = P\{N(2) = 0\} = e^{-\lambda t} = e^{-2 \cdot 2} \doteq 0.018,$$

$$P_1(2) = P\{N(2) = 1\} = \lambda t e^{-\lambda t} = 2 \cdot 2 e^{-2 \cdot 2} \doteq 0.073,$$

$$P_2(2) = P\{N(2) = 2\} = \frac{(\lambda t)^2}{2!} e^{-\lambda t} = \frac{(2 \cdot 2)^2}{2} e^{-2 \cdot 2} \doteq 0.147.$$

(2) Calculate the mean and variance of the customers arriving during business hours. Noting that $t = 8$ (from 10 *a.m.* to 6 *p.m.*), we have

$$E[N(8)] = \lambda t = 2 \cdot 8 = 16 \text{ persons},$$

and

$$Var(N(8)) = \lambda t = 2 \cdot 8 = 16 \text{ persons}^2.$$

The Poisson process is a counting process $\{N(t),\ t \geq 0\}$ with $N(t) \sim POI(\lambda t)$. We show several additional, interesting properties of the Poisson process in the following section.

3.3 Interarrival Time Distributions

In the preceding section we have discussed the Poisson process focusing on the random variable $N(t)$. However, in this section we focus on the time intervals in which two or more consecutive events take place.

Definition 3.3.1 Consider a counting process $\{N(t),\ t \geq 0\}$. Let X_1 be the time of the first event or occurrence. In general, let X_n be the time between $(n-1)$st and nth events. Then $\{X_n,\ n = 1, 2, \cdots\}$ is called the sequence of *interarrival* or *interoccurrence times.* Further, let S_n be the *waiting* or *arriving time* of the nth event, i.e.,

$$S_n = X_1 + X_2 + \cdots + X_n \qquad (n = 1, 2, \cdots), \tag{3.3.1}$$

where $S_0 = 0$.

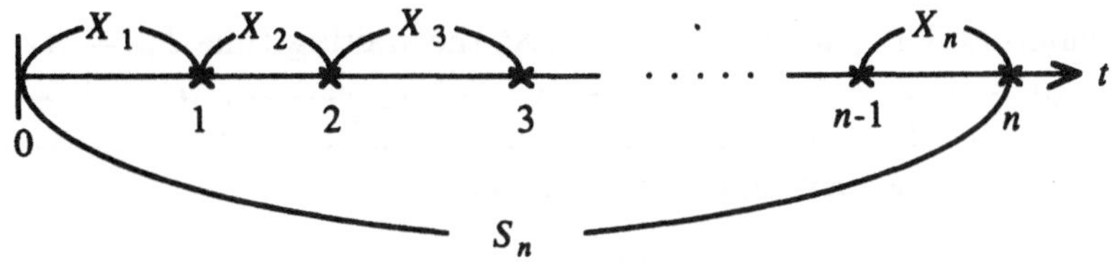

Fig. 3.3.1 A realization of interarrival and waiting times.

Fig. 3.3.1 shows a realization of interarrival and waiting times. We first show the following theorem.

Theorem 3.3.1 For a Poisson process with parameter $\lambda > 0$, the interarrival times X_n $(n = 1, 2, \cdots)$ are independent and identically distributed exponential random variables with mean $1/\lambda$.

Proof It is easy to show that

$$P\{X_1 \le t\} = 1 - P\{X_1 > t\} = 1 - P\{N(t) = 0\} = 1 - e^{-\lambda t}, \qquad (3.3.2)$$

i.e., $X_1 \sim EXP(\lambda)$. For X_2, we have the conditional distribution that the first event takes place at time s:

$$\begin{aligned} & P\{X_2 \le t | X_1 = s\} \\ &= 1 - P\{X_2 > t | X_1 = s\} \\ &= 1 - P\{N(t+s) - N(s) = 0 | X_1 = s\} \\ &= 1 - P\{N(t+s) - N(s) = 0\} && \text{(from independent increments)} \\ &= 1 - P\{N(t) = 0\} && \text{(from stationary increments)} \\ &= 1 - e^{-\lambda t}, && (3.3.3) \end{aligned}$$

i.e., the conditional distribution $P\{X_2 \le t | X_1 = s\}$ is independent of X_1 and $X_2 \sim EXP(\lambda)$. We can recursively show that each interarrival time X_n is independent and identically distributed exponentially, which proves the theorem.

In the following chapter we will introduce a *renewal process* which is roughly defined as a counting process for which the interarrival times are independent and identically distributed with say, an arbitrary distribution $F(t)$. We are ready to show an alternative definition of the Poisson process.

Definition 3.3.2 A Poisson process with parameter $\lambda > 0$ is a renewal process with the exponential interarrival distribution $F(t) = 1 - e^{-\lambda t}$.

It is obvious that Definitions 3.2.2, 3.2.3 and 3.3.2 are equivalent. For a Poisson process we can understand that $N(t) \sim POI(\lambda t)$ and $X_k \sim EXP(\lambda)$ ($k = 1, 2, \cdots$). We further show the other distributions derived from a Poisson process.

In Definition 3.3.1, we have introduced the waiting time $S_n = X_1 + X_2 + \cdots + X_n$ ($n = 1, 2, \cdots$), where $S_0 = 0$. Noting that X_n ($n = 1, 2, \cdots$) are independent and identically distributed exponential random variables, we have $S_n \sim GAM(\lambda, n)$ (see Example 2.5.4). That is,

$$P\{S_n \le t\} = \int_0^t \frac{\lambda(\lambda x)^{n-1} e^{-\lambda x}}{(n-1)!} dx. \tag{3.3.4}$$

Let us recall the relationship between S_n and $N(t)$ (see Fig. 3.3.2). It is obvious that

$$S_n \le t \Leftrightarrow N(t) \ge n. \tag{3.3.5}$$

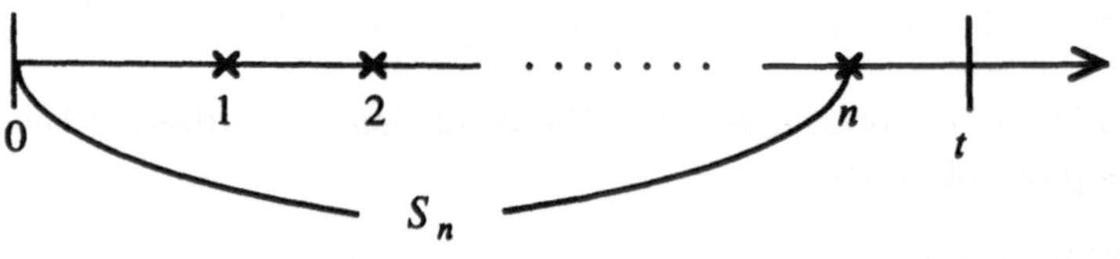

Fig. 3.3.2 The relationship between $S_n \le t$ and $N(t) \ge n$.

Thus the statement that the waiting time of the nth event is less than or equal to t is equivalent to the statement that the number of events up to time t is greater than or equal to n. Referring to Eqs.(3.3.4) and (3.3.5), we have the following:

Theorem 3.3.2 For a Poisson process with parameter $\lambda > 0$, we have

$$P\{S_n \le t\} = P\{N(t) \ge n\}, \tag{3.3.6}$$

i.e.,

$$\int_0^t \frac{\lambda(\lambda x)^{n-1} e^{-\lambda x}}{(n-1)!} dx = \sum_{i=n}^{\infty} \frac{(\lambda t)^i}{i!} e^{-\lambda t}. \tag{3.3.7}$$

Equivalently, we have

$$P\{S_n > t\} = P\{N(t) < n\}, \tag{3.3.8}$$

i.e.,

$$\int_t^{\infty} \frac{\lambda(\lambda x)^{n-1}e^{-\lambda x}}{(n-1)!}dx = \sum_{i=0}^{n-1} \frac{(\lambda t)^i}{i!}e^{-\lambda t}. \tag{3.3.9}$$

The identity (3.3.7) or (3.3.9) can be alternatively verified from the analysis of applying n iterations of integration by parts (see Problem 3.9). However, we can derive the identity (3.3.7) or (3.3.9) directly from the probabilistic interpretation of the Poisson process without this cumbersome analysis.

3.4 Conditional Waiting Time Distributions

Consider the conditional distribution of the first arrival time X_1, given that there was an event in the time interval $[0, t]$. That is, for $s \leq t$

$$P\{X_1 \leq s | N(t) = 1\} = \frac{P\{N(s) = 1,\ N(t) - N(s) = 0\}}{P\{N(t) = 1\}}$$

$$= \frac{\lambda s e^{-\lambda s} \cdot e^{-\lambda(t-s)}}{\lambda t e^{-\lambda t}} = \frac{s}{t}, \tag{3.4.1}$$

which is a uniform distribution, i.e., $X_1 \sim U(0, t)$. That is, the probability that an event occurs given that there was an event in the time interval $[0, t]$ is uniformly distributed over $[0, t]$ (see Fig. 3.4.1). The discrete version of this fact was shown in Section 3.1 (i.e., the (discrete) first arrival time is uniformly distributed). To generalize this fact, we can show the following:

Theorem 3.4.1 The conditional distribution of n waiting times $S_1,\ S_2,\ \cdots,\ S_n$ given that $N(t) = n$ is

$$P\{S_1 \leq s_1,\ S_2 \leq s_2,\ \cdots,\ S_n \leq s_n | N(t) = n\}$$

$$= n! \int_0^{s_1} \int_{s_1}^{s_2} \cdots \int_{s_{n-1}}^{s_n} \frac{1}{t^n} dx_1\, dx_2 \cdots dx_n. \tag{3.4.2}$$

That is, the conditional distribution (3.4.2) is the same as the order statistics corresponding to n independent random variables uniformly distributed over the interval $[0, t]$.

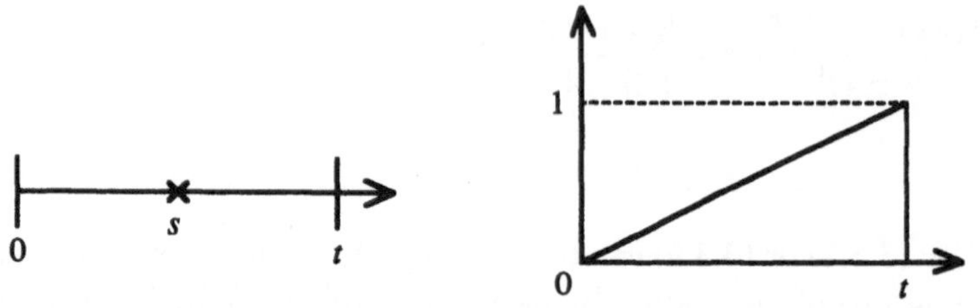

(a) a realization (b) $P\{X_1 \leq s \mid N(t) = 1\}$

Fig. 3.4.1 Conditional waiting time distribution $P\{X_1 \leq s \mid N(t) = 1\}$.

The conditional density of n waiting times $S_1, S_2, \cdots, S_n$ given that $N(t) = n$ is

$$f(t_1, t_2, \cdots, t_n | N(t) = n) = \frac{n!}{t^n} \qquad (0 < t_1 < t_2 < \cdots < t_n), \quad (3.4.3)$$

which can be interpreted to mean that unordered random variables of n waiting times $S_1, S_2, \cdots, S_n$, given that $N(t) = n$, are independent and identically distributed uniformly over the interval $[0, t]$.

Theorem 3.4.1 can frequently be used for statistical inferences under the assumptions of exponential lifetime distributions in reliability theory. Theorem 3.4.1 can be directly applied to several probability models arising in queueing theory and inventory control.

Example 3.4.1 (*Infinite server Poisson queue*) Consider a Poisson queue, i.e., the law of probability of the consecutive customers' arrivals follows a Poisson process with parameter λ. The customers who have arrived are served immediately with a common and arbitrary distribution $G(t)$. Then, what is the probability that there are k customers served at time $t \geq 0$? Assume that there are n customers who have arrived during the time interval t, which implies that unordered random variables of n arrivals $S_1, S_2, \cdots, S_n$ are independent and identically distributed uniformly over the interval $[0, t]$. Referring to Fig. 3.4.2, the probability that a customer is being served at time t is

$$p = \int_0^t [1 - G(t - x)] \frac{dx}{t} = \frac{1}{t} \int_0^t [1 - G(x)] dx.$$

Thus, the probability that there are k customers being served is

$$P\{k \text{ customers being served at time } t\}$$

$$= \sum_{n=k}^{\infty} \binom{n}{k} p^k (1-p)^{n-k} \frac{(\lambda t)^n}{n!} e^{-\lambda t} = \frac{(p\lambda t)^k}{k!} e^{-p\lambda t} \qquad (k = 0, 1, 2, \cdots),$$

i.e., the probability that k customers are being served at time t follows a Poisson distribution with mean

$$p\lambda t = \lambda \int_0^t [1 - G(x)]dx,$$

which is a function of t (see the Poisson distribution in Section 2.3).

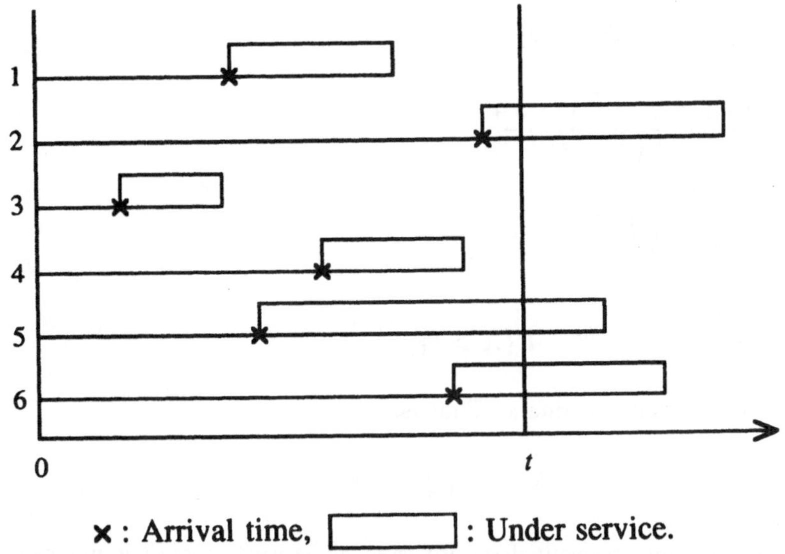

Fig. 3.4.2 A realization of 3 customers being served given that 6 customers arrived over $[0, t]$.

Let us introduce the following random variables:

$$\delta_t = t - S_{N(t)} \qquad (\textit{current life} \text{ or } \textit{age}), \tag{3.4.4}$$

$$\gamma_t = S_{N(t)+1} - t \qquad (\textit{excess life} \text{ or } \textit{residual life}), \tag{3.4.5}$$

$$\begin{aligned} \beta_t &= \delta_t + \gamma_t \\ &= S_{N(t)+1} - S_{N(t)} \\ &= X_{N(T)+1} \qquad (\textit{total life}). \end{aligned} \tag{3.4.6}$$

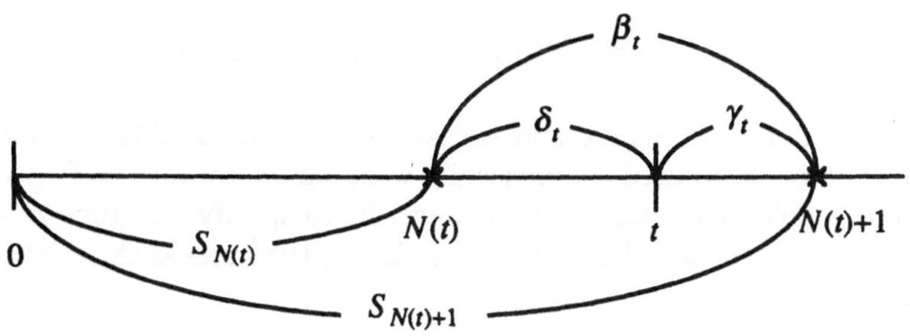

Fig. 3.4.3 Realizations of the current life (age) δ_t, excess life (residual life) γ_t, and total life β_t.

Figure 3.4.3 shows the meanings of the random variables above. It is clear that total life is distributed exponentially with parameter λ.

In general, let X be a random variable with the exponential distribution $F(x) = 1 - e^{-\lambda x}$. Then the following conditional survival probability is given by

$$\begin{aligned} & P\{X > s+t | X > s\} \\ &= \frac{P\{X > s \text{ and } X > s+t\}}{P\{X > s\}} \\ &= \frac{P\{X > s+t\}}{P\{X > s\}} \\ &= \frac{e^{-\lambda(s+t)}}{e^{-\lambda s}} = e^{-\lambda t} = P\{X > t\}, \end{aligned} \tag{3.4.7}$$

which is independent of time s. That is,

$$P\{X > s+t\} = P\{X > s\} P\{X > t\}. \tag{3.4.8}$$

The condition that an event does not take place up to time t is not required to derive the conditional probability (3.4.7). Such a property in Eq.(3.4.7) or (3.4.8) is called the *memoryless property*. Conversely, for all $s > 0$ and $t > 0$, the only function which satisfies the functional equation in Eq.(3.4.8) is $P\{X > t\} = e^{-\lambda t}$. That is, only the exponential distribution has the memoryless property.

Noting that the Poisson process has stationary independent increments and only the exponential distribution has the memoryless property, we have

$$P\{\delta_t > x\} = P\{\gamma_t > x\} = e^{-\lambda x}. \tag{3.4.9}$$

and the joint probability that $\delta_t > x$ and $\gamma_t > y$

$$P\{\gamma_t > x,\ \delta_t > y\} = e^{-\lambda(x+y)} \qquad (x > 0, 0 < y < t). \tag{3.4.10}$$

That is, both the current life and excess life random variables are independent and identically distributed exponentially with parameter λ. The Poisson process has quite simple properties: For instance, if we specify any time $t > 0$ as the time origin, the resultant Poisson process is just the same as the original Poisson process.

We discuss the superposition and decomposition of Poisson processes. First, we discuss the superposition of two independent Poisson processes (see Fig.

3.4.4).

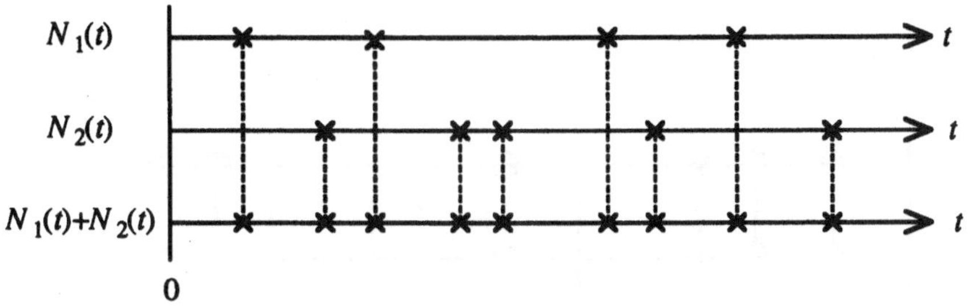

Fig. 3.4.4 A realization of superposition.

Theorem 3.4.2 (*Superposition of the Poisson Processes*) The pooled process $\{N_1(t) + N_2(t), t \geq 0\}$ of two independent Poisson processes $\{N_1(t),\ t \geq 0\}$ and $\{N_2(t),\ t \geq 0\}$ with respective parameters λ and μ is again a Poisson process with parameter $\lambda + \mu$.

Proof

$$P\{N_1(t) + N_2(t) = n\}$$

$$= \sum_{k=0}^{n} P\{N_1(t) = k,\ N_2(t) = n-k\}$$

$$= \sum_{k=0}^{n} \frac{(\lambda t)^k}{k!} e^{-\lambda t} \frac{(\mu t)^{n-k}}{(n-k)!} e^{-\mu t}$$

$$= \frac{[(\lambda+\mu)t]^n}{n!} e^{-(\lambda+\mu)t} \qquad (n = 0,\ 1,\ 2,\ \cdots), \tag{3.4.11}$$

which is the probability mass function of a Poisson distribution with parameter $(\lambda + \mu)t$ and which completes the proof.

Secondly, we discuss the decomposition of a Poisson process (see Fig. 3.4.5). Let us consider a Poisson process in which the consecutive events follow the Bernoulli trials. For instance, road traffic flow can be described by a Poisson process in which two categories of cars (e.g., passenger vehicles and non-passenger vehicles) can be classified. We are ready to show the following:

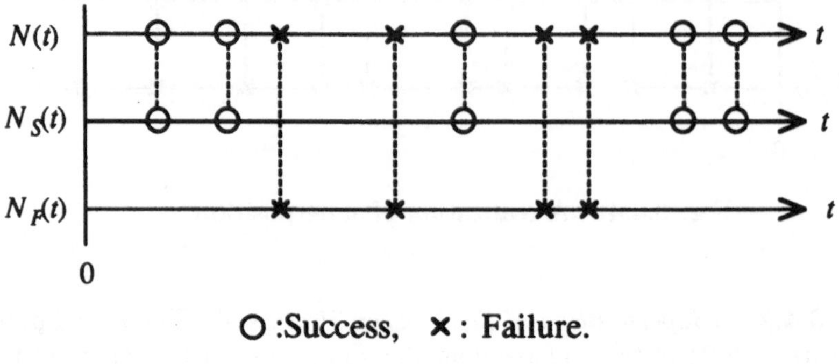

Fig. 3.4.5 A realization of decomposition.

Theorem 3.4.3 (*Decomposition of a Poisson Process*) Let $\{N(t),\ t \geq 0\}$ be a Poisson process with a parameter λ and let the consecutive events follow the Bernoulli trials with parameter p (the probability of success). The classified processes $\{N_S(t),\ t \geq 0\}$ and $\{N_F(t),\ t \geq 0\}$ of success and failure are independent Poisson processes with respective parameters $p\lambda$ and $q\lambda$, where $q = 1 - p$ (the probability of failure).

Proof The joint probability is given by

$$\begin{aligned} &P\{N_S(t) = k,\ N_F(t) = n - k\} \\ &= P\{N_S(t) = k,\ N_F(t) = n - k | N(t) = n\} P\{N(t) = n\} \\ &= \binom{n}{k} p^k q^{n-k} \frac{(\lambda t)^n}{n!} e^{-\lambda t} \\ &= \frac{(p\lambda t)^k}{k!} e^{-p\lambda t} \cdot \frac{(q\lambda t)^{n-k}}{(n-k)!} e^{-q\lambda t}, \end{aligned} \tag{3.4.12}$$

which implies that $\{N_S(t),\ t \geq 0\}$ and $\{N_F(t),\ t \geq 0\}$ are independent Poisson processes with respective parameters $p\lambda$ and $q\lambda$. It is easy to generalize Theorems 3.4.2 and 3.4.3 with n Poisson processes.

Example 3.4.2 Let $\{N(t),\ t \geq 0\}$ be a Poisson process with parameter λ describing road traffic flow. The passing vehicles can be classified in three categories: The probabilities of A, B, and C (say, A: passenger car, B: bus, C: truck) are p, q, and r, respectively, where $p + q + r = 1$ and each vehicle is independent of the others. Using the multinomial coefficient (see Theorem 1.4.5), the joint probability is given by

$$P\{N_A(t) = k,\ N_B(t) = l,\ N_C(t) = m\}$$

$$= P\{N_A(t) = k,\ N_B(t) = l,\ N_C(t) = m | N(t) = k + l + m\}$$
$$\times P\{N(t) = k + l + m\}$$
$$= \binom{k+l+m}{k \quad l \quad m} p^k q^l r^m \frac{(\lambda t)^{k+l+m}}{(k+l+m)!} e^{-\lambda t}$$
$$= \frac{(p\lambda t)^k}{k!} e^{-p\lambda t} \cdot \frac{(q\lambda t)^l}{l!} e^{-q\lambda t} \cdot \frac{(r\lambda t)^m}{m!} e^{-r\lambda t}$$
$$= P\{N_A(t) = k\} P\{N_B(t) = l\} P\{N_C(t) = m\},$$

which implies that the three processes are independent Poisson processes with respective parameters $p\lambda$, $q\lambda$, and $r\lambda$.

3.5 Nonhomogeneous Poisson Processes

We have assumed that the Poisson process has stationary independent increments. Thus the Poisson process should be called the *homogeneous* or *stationary Poisson process.* In this section, eliminating the stationarity we generalize the Poisson process with the parameter which is a function of time t.

Definition 3.5.1 A counting process $\{N(t),\ t \geq 0\}$ is called a *nonhomogeneous* or *nonstationary Poisson process* with the *intensity function* $\lambda(t)$ if the following conditions are satisfied:

(i) $N(0) = 0$.

(ii) The process has independent increments.

(iii) $P\{N(t+h) - N(t) = 1\} = \lambda(t)h + o(h)$.

(iv) $P\{N(t+h) - N(t) \geq 2\} = o(h)$.

The homogeneous Poisson process has a constant parameter λ. However, the nonhomogeneous Poisson process has a function $\lambda(t)$ which is called the *intensity function.* Let

$$m(t) = \int_0^t \lambda(x)dx \tag{3.5.1}$$

be the integral of $\lambda(t)$ from 0 to t. Then we can similarly derive the probability

$$P_k(t) = P\{N(t) = k | N(0) = 0\}$$
$$= \frac{[m(t)]^k}{k!} e^{-m(t)}, \tag{3.5.2}$$

where $E[N(t)] = m(t)$ is the mean for a fixed t and is called the *mean value function*. Noting the independence of the process, we can show that

$$P\{N(t+s) - N(t) = k\} = \frac{[m(t+s) - m(t)]^k}{k!} e^{-[m(t+s)-m(t)]}, \quad (3.5.3)$$

where

$$m(t+s) - m(t) = \int_t^{t+s} \lambda(x)dx. \quad (3.5.4)$$

Similar to Theorem 3.3.2 for the homogeneous Poisson process, we can show that

$$S_n \leq t \Leftrightarrow N(t) \geq n \quad (3.5.5)$$

for the nonhomogeneous Poisson process. That is

$$P\{S_n \leq t\} = P\{N(t) \geq n\}, \quad (3.5.6)$$

or

$$\int_0^t \frac{\lambda(x)[m(x)]^{n-1}}{(n-1)!} e^{-m(x)} dx = \sum_{i=n}^{\infty} \frac{[m(t)]^i}{i!} e^{-m(t)}. \quad (3.5.7)$$

Of course, it is possible to verify the identity (3.5.7) by the analysis of applying n iterations of integration by parts. The verification by probabilistic interpretation is simpler than that of analysis.

The conditional distribution of the first arrival time S_1 given that $N(t) = 1$ is

$$\begin{aligned} & P\{S_1 \leq s | N(t) = 1\} \\ &= \frac{P\{S_1 \leq s,\ N(t) = 1\}}{P\{N(t) = 1\}} \\ &= \frac{P\{N(s) = 1,\ N(t) - N(s) = 0\}}{P\{N(t) = 1\}} \\ &= \frac{m(s)e^{-m(s)}e^{-[m(t)-m(s)]}}{m(t)e^{-m(t)}} = \frac{m(s)}{m(t)} \qquad (s \leq t). \end{aligned} \quad (3.5.8)$$

Generalizing this fact, we have the following theorem which corresponds to Theorem 3.4.1 for the nonhomogeneous Poisson process.

Theorem 3.5.1 The conditional distribution of n waiting times $S_1,\ S_2,\ \cdots,\ S_n$ given that $N(t) = n$ is

$$P\{S_1 \leq s_1,\ S_2 \leq s_2,\ \cdots,\ S_n \leq s_n, | N(t) = n\}$$

$$= n! \int_0^{s_1} \int_{s_1}^{s_2} \cdots \int_{s_{n-1}}^{s_n} \frac{\prod_{i=1}^{n} \lambda(x_i)}{[m(t)]^n} dx_1\, dx_2 \cdots dx_n. \tag{3.5.9}$$

The conditional density of n waiting times $S_1, S_2, \cdots, S_n$ given that $N(t) = n$ is

$$f(t_1, t_2, \cdots, t_n | N(t) = n) = \frac{n!}{[m(t)]^n} \prod_{i=1}^{n} \lambda(t_i), \tag{3.5.10}$$

which can be interpreted to mean that unordered random variables of n waiting times $S_1, S_2, \cdots, S_n$ given that $N(t) = n$ are independent and identically distributed with the density

$$f(x) = \begin{cases} \frac{\lambda(x)}{m(t)} & (0 \le x \le t) \\ 0 & (\text{otherwise}). \end{cases} \tag{3.5.11}$$

In particular, $m(t) = \lambda t$ (i.e., $\lambda(t) = \lambda$) implies that $f(x)$ is uniformly distributed over $[0, t]$, which is a special case of the nonhomogeneous Poisson process, given by Theorem 3.4.1.

3.6 Problems 3

3.1 Verify Eq.(3.2.10) by the techniques of mathematical induction.

3.2 Let

$$g(t, s) = \sum_{n=0}^{\infty} P_n(t)s^n \qquad (|s| < 1)$$

be the generating function of $P_n(t)$. By multiplying Eqs.(3.2.8) and (3.2.9) by s^n and summing over n, show that the generating function satisfies

$$\frac{\partial g(t, s)}{\partial t} = \lambda(s-1)g(t, s).$$

Solve for $g(t, s)$ and derive $P_n(t)$ by expanding as a power series of n.

3.3 For a shop opening from 9 *a.m.* to 6 *p.m.*, the arriving customers obey a Poisson process with 10 persons per hour throughout business hours.

(i) Calculate the mean and variance of the arriving customers during business hours.

(ii) Calculate the probability that no customers arrive for half an hour.

3.4 For a shop opening from 9 *a.m.* to 6 *p.m.*, the arriving customers obey a Poisson process with a mean interarrival time of 6 minutes.

(i) Calculate the probabilities that k customers ($k = 0, 1, 2$) arrive within half an hour.

(ii) Calculate the mean and variance of the arriving customers during business hours, and derive the lower and upper number of arriving customers for the mean $\pm 3\sqrt{Var}$.

3.5 Let $\{N(t),\ t \geq 0\}$ be a Poisson process. Show that

$$P\{N(s) = k | N(t) = n\} = \binom{n}{k}\left(\frac{s}{t}\right)^k\left(1-\frac{s}{t}\right)^{n-k} \qquad (k = 0, 1, \cdots, n)$$

for $s < t$.

3.6 (*Continuation*) Verify Problem 3.5 by applying Theorem 3.4.1.

3.7 Let $\{N_1(t),\ t \geq 0\}$ and $\{N_2(t),\ t \geq 0\}$ be independent Poisson processes with respective parameters λ_1 and λ_2. Consider the combined process $\{N_1(t) + N_2(t),\ t \geq 0\}$ by superposition. Show that the probability that the first event of the combined process takes place from $\{N_1(t),\ t \geq 0\}$ is $\lambda_1/(\lambda_1 + \lambda_2)$ and is independent of the time of the event.

3.8 Let $\{N(t),\ t \geq 0\}$ be a Poisson process with parameter λ. Calculate $E[N(t) \cdot N(t+s)]$.

Hint: Calculate $E[N(t)\{N(t+s) - N(t)\}]$ by using the stationarity of the Poisson process.

3.9 Verify the identity (3.3.9) by the following analysis of integration by parts: Define

$$I_n(t) = \int_t^\infty \frac{\lambda(\lambda x)^{n-1}e^{-\lambda x}}{(n-1)!}dx$$

$$= -\frac{(\lambda x)^{n-1}}{(n-1)!}e^{-\lambda x}\bigg|_t^\infty + \int_t^\infty \frac{\lambda(\lambda x)^{n-2}e^{-\lambda x}}{(n-2)!}dx$$

$$= \frac{(\lambda t)^{n-1}}{(n-1)!}e^{-\lambda t} + I_{n-1}(t)$$

and repeat the similar integration by parts.

3.10 A supermarket opens at 10 *a.m.* ($t = 0$) and closes at 8 *p.m.* ($t = 10$). The arriving customers obey a nonhomogeneous Poisson process with the following intensity function:

$$\lambda(t) = \begin{cases} 20t & (0 \leq t \leq 2) \\ 10t + 20 & (2 \leq t \leq 6) \\ 80 & (6 \leq t \leq 8) \\ 400 - 40t & (8 \leq t \leq 10). \end{cases}$$

(i) Calculate the mean and variance of the arriving customers during business hours from 10 *a.m.* to 8 *p.m.*

(ii) Calculate the mean for arriving customers during the busiest period from 4 *p.m.* to 6 *p.m.*

3.11 Consider a nonhomogeneous Poisson process in which the intensity function is given by $\lambda(t) = \alpha\beta(\alpha t)^{\beta-1}$. Derive the distribution of the first interarrival time.

3.12 Consider a nonhomogeneous Poisson process $\{N(t), t \geq 0\}$ with the intensity function $\lambda(t)$. Let

$$m(t) = \int_0^t \lambda(x)dx$$

be the mean value function of the process. Recall that $\{X_n, n = 1, 2, \cdots\}$ is the sequence of interarrival times of the process.

(i) Show that

$$P\{X_1 \leq x\} = 1 - e^{-m(x)},$$

and

$$P\{X_n \leq x\} = 1 - \int_0^\infty e^{-m(t+x)} \frac{[m(t)]^{n-2}}{(n-2)!} \lambda(t) dt$$

$$(n = 2, 3, \cdots).$$

(ii) Show that

$$E[X_n] = \int_0^\infty \frac{[m(x)]^{n-1}}{(n-1)!} e^{-m(x)} dx$$

$$(n = 1, 2, 3, \cdots).$$

Chapter 4

Renewal Processes

4.1 Introduction

In the preceding chapter we have introduced the Poisson process from two different viewpoints. According to Definition 3.2.2 the Poisson process has stationary independent increments and the probability that an event takes place for a small interval h is λh, where the proportional constant λ is the parameter of the process. Furthermore, according to Definition 3.3.2 the Poisson process is a renewal process in which the interarrival times are independent and identically distributed exponentially with the mean $1/\lambda$. Figure 4.1.1 shows two realizations of the Poisson process from these two viewpoints.

It is quite natural to generalize the Poisson process by allowing an arbitrary interarrival time distribution. We are now in a position to define a renewal process.

Definition 4.1.1 A counting process $\{N(t),\ t \geq 0\}$ is called a *renewal process* if the interarrival times are independent and identically distributed with an arbitrary distribution $F(t)$.

In this chapter we are concerned with the renewal process and discuss its interesting properties. There are many applications of the renewal process in science, engineering, business, and social sciences.

Throughout this chapter the Poisson process is discussed frequently as a special case of the renewal processes. Of course, the Poisson process is one of the simplest examples of renewal processes and gives us interesting insights into the renewal process because of its simplicity.

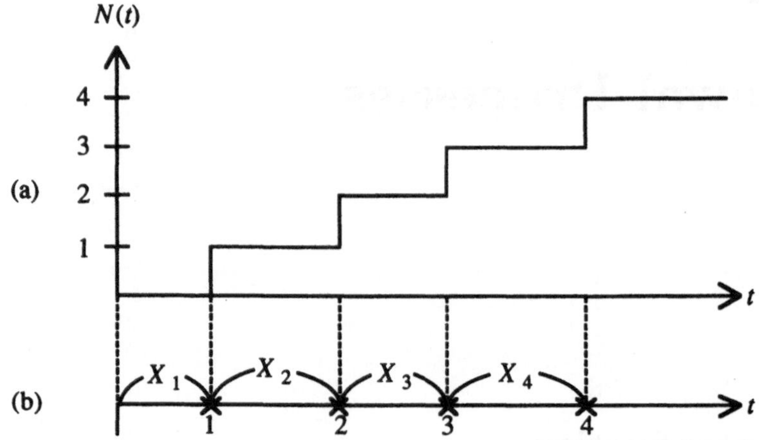

Fig. 4.1.1 Two realizations of the Poisson process.

4.2 Renewal Functions

For a renewal process in Definition 4.1.1, we defined an identical interarrival time distribution which is arbitrary in general:

$$F(t) = P\{X_m \le t\} \qquad (m = 1,\ 2,\ \cdots). \tag{4.2.1}$$

Note that the random variable X_m is *non-negative* for the renewal process. Let S_n be the waiting time of the nth event, i.e.,

$$S_n = X_1 + X_2 + \cdots + X_n \qquad (n = 1,\ 2,\ \cdots), \tag{4.2.2}$$

where we postulate $S_0 = 0$ for convenience.

As shown in Section 2.5, we have the distribution $P\{S_n \le t\}$ of the nth waiting time S_n:

$$P\{S_n \le t\} = F^{(n)}(t) = \overbrace{F * F * \cdots * F(t)}^{n} \qquad (n = 1,\ 2,\ \cdots), \tag{4.2.3}$$

which is the n-fold Stieltjes convolution of the interarrival distribution $F(t)$ of itself. It is convenient for us to define

$$\begin{aligned} F^{(n)}(t) &= F * F^{(n-1)}(t) \\ &= \int_0^t F^{(n-1)}(t - x) dF(x) \qquad (n = 1,\ 2,\ 3,\ \cdots), \end{aligned} \tag{4.2.4}$$

with $F^{(0)}(t) = 1(t)$ (a step function) and $F^{(1)} = F(t)$.

Referring to Fig. 3.3.2 for a counting process, we have the following relationship:

$$S_n \leq t \iff N(t) \geq n, \tag{4.2.5}$$

which implies

$$P\{N(t) \geq n\} = P\{S_n \leq t\} = F^{(n)}(t), \tag{4.2.6}$$

and

$$\begin{aligned} &P\{N(t) = n\} \\ &= P\{N(t) \geq n\} - P\{N(t) \geq n+1\} \\ &= F^{(n)}(t) - F^{(n+1)}(t). \end{aligned} \tag{4.2.7}$$

The mean of the random variable $N(t)$ is called the *renewal function* $M(t) = E[N(t)]$, which is the mean number of *renewals* or *events* up to time t and is given by

$$\begin{aligned} M(t) &= E[N(t)] \\ &= \sum_{m=1}^{\infty} mP\{N(t) = m\} \\ &= \sum_{m=1}^{\infty} \sum_{n=1}^{m} P\{N(t) = m\} \qquad (\sum_{n=1}^{m} 1 = m) \\ &= \sum_{n=1}^{\infty} \sum_{m=n}^{\infty} P\{N(t) = m\} \\ &= \sum_{n=1}^{\infty} P\{N(t) \geq n\} \\ &= \sum_{n=1}^{\infty} F^{(n)}(t), \end{aligned} \tag{4.2.8}$$

where we applied a formula of interchanging the orders of summation:

$$\sum_{m=1}^{\infty} \sum_{n=1}^{m} a_{mn} = \sum_{n=1}^{\infty} \sum_{m=n}^{\infty} a_{mn}, \tag{4.2.9}$$

(see Problem 4.1). That is, the renewal function is a series of the n-fold Stieltjes convolutions of $F(t)$ of itself.

Example 4.2.1 If we assume $F(t) = 1 - e^{-\lambda t}$ for a renewal process, we have

$$F^{(n)}(t) = \int_0^t \frac{\lambda(\lambda x)^{n-1}e^{-\lambda x}}{(n-1)!}dx = \sum_{i=n}^{\infty} \frac{(\lambda t)^i}{i!}e^{-\lambda t},$$

$$P\{N(t) = n\} = F^{(n)}(t) - F^{(n+1)}(t) = \frac{(\lambda t)^n}{n!}e^{-\lambda t},$$

and

$$\begin{aligned} M(t) &= E[N(t)] \\ &= \sum_{n=1}^{\infty} \int_0^t \frac{\lambda(\lambda x)^{n-1}}{(n-1)!}e^{-\lambda x}dx \\ &= \lambda \int_0^t e^{-\lambda x} \sum_{n=1}^{\infty} \frac{(\lambda x)^{n-1}}{(n-1)!}dx \\ &= \lambda \int_0^t dx = \lambda t, \end{aligned}$$

where the interchange of integral and summation is justified since all the terms are positive and bounded. Of course, the results in this example were examined in Section 3.2 as the results of the Poisson process.

Example 4.2.2 If we assume that

$$F(t) = \int_0^t \lambda(\lambda x)e^{-\lambda x}dx = 1 - (1 + \lambda t)e^{-\lambda x},$$

i.e., $X_m \sim GAM(\lambda,\ 2)$ $(m = 1,\ 2,\ \cdots)$ (ref. Section 2.4 (iii)), we have

$$P\{N(t) = n\} = \sum_{i=2n}^{2n+1} \frac{(\lambda t)^i}{i!}e^{-\lambda t} \qquad (n = 0,\ 1,\ 2,\ \cdots).$$

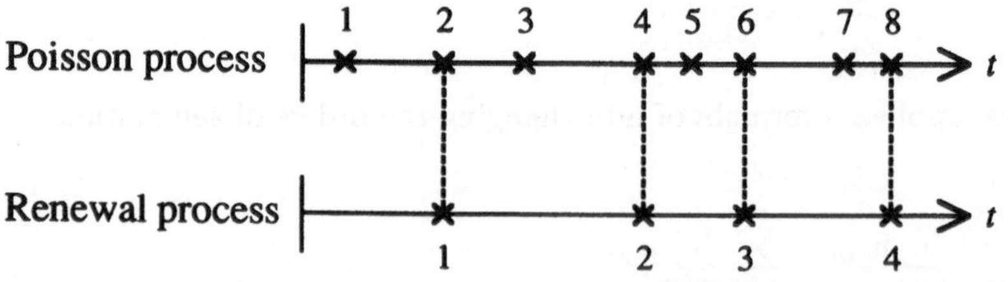

Fig. 4.2.1 A relationship between a renewal process with $X_m \sim GAM(\lambda,\ 2)$ and a Poisson process with parameter λ.

Referring to Fig. 4.2.1, we can understand that any two consecutive events for the Poisson process make an event for the renewal process since $X_m \sim GAM(\lambda,\ 2)$ implies that the distribution of the sum of two independent exponential random variables is a gamma distribution of order 2. The renewal function $M(t)$ is given by

$$M(t) = \sum_{k=1}^{\infty} k \left[\frac{(\lambda t)^{2k}}{(2k)!} e^{-\lambda t} + \frac{(\lambda t)^{2k+1}}{(2k+1)!} e^{-\lambda t} \right]$$

$$= \frac{\lambda t}{2} - \frac{1}{4} + \frac{1}{4} e^{-2\lambda t},$$

(see Problem 4.2).

Example 4.2.3 (*Continuation*) It is plausible to generalize Example 4.2.2 by allowing k in general, where k is a positive integer. If we assume that

$$F(t) = \int_0^t \frac{\lambda(\lambda x)^{k-1} e^{-\lambda x}}{(k-1)!} dx = \sum_{i=k}^{\infty} \frac{(\lambda t)^i}{i!} e^{-\lambda t},$$

i.e., $X_m \sim GAM(\lambda,\ k)$ $(m = 1,\ 2,\ \cdots)$ (ref. Section 2.4 (iii)), we have

$$P\{N(t) = n\} = \sum_{i=nk}^{nk+k-1} \frac{(\lambda t)^i}{i!} e^{-\lambda t},$$

and

$$M(t) = \frac{\lambda t}{k} + \frac{1}{k} \sum_{r=1}^{k-1} \frac{\varepsilon_r}{1-\varepsilon_r} \left[1 - e^{-\lambda t(1-\varepsilon_r)} \right],$$

where $\varepsilon_r = e^{\frac{2\pi r i}{k}}$ $(r = 0,\ 1,\ 2,\ \cdots, k-1)$ are the distinct roots of the equation $s^k = 1$.

The renewal function $M(t)$ can be rewritten as

$$M(t) = \sum_{n=1}^{\infty} F^{(n)}(t)$$

$$= F(t) + \sum_{n=1}^{\infty} F * F^{(n)}(t)$$

$$= F(t) + F * \sum_{n=1}^{\infty} F^{(n)}(t)$$

$$= F(t) + F * M(t)$$
$$= F(t) + \int_0^t M(t-x)dF(x), \tag{4.2.10}$$

where the interchange of integral and summation is justified. That is,

$$M(t) = F(t) + \int_0^t M(t-x)dF(x) \tag{4.2.11}$$

is called the *renewal equation* which is an integral equation with known $F(t)$ and unknown $M(t)$.

Noting the convolution form in Eq.(4.2.11), we introduce the Laplace-Stieltjes transforms:

$$F^*(s) = \int_0^\infty e^{-st}dF(t), \tag{4.2.12}$$

and

$$M^*(s) = \int_0^\infty e^{-st}dM(t). \tag{4.2.13}$$

Taking the Laplace-Stieltjes transforms for Eq.(4.2.11), we have

$$M^*(s) = F^*(s) + F^*(s)M^*(s), \tag{4.2.14}$$

and

$$M^*(s) = \frac{F^*(s)}{1 - F^*(s)}. \tag{4.2.15}$$

Equation (4.2.15) asserts that $F^*(s)$ (i.e., $F(t)$) uniquely specifies $M^*(s)$ (i.e., $M(t)$), and vice versa.

Example 4.2.4 (*Continuation of Example 4.2.1*) The Laplace-Stieltjes transform of $F(t) = 1 - e^{-\lambda t}$ is given by

$$F^*(s) = \frac{\lambda}{s+\lambda},$$

and

$$M^*(s) = \frac{\frac{\lambda}{s+\lambda}}{1 - \frac{\lambda}{s+\lambda}} = \frac{\lambda}{s},$$

which implies $M(t) = \lambda t$ by inversion.

Example 4.2.5 (*Continuation of Example 4.2.2*) The Laplace-Stieltjes transform of $F(t) = 1 - (1 + \lambda t)e^{-\lambda t}$ is given by

$$F^*(s) = \left(\frac{\lambda}{s+\lambda}\right)^2,$$

and

$$M^*(s) = \frac{\left(\frac{\lambda}{s+\lambda}\right)^2}{1 - \left(\frac{\lambda}{s+\lambda}\right)^2} = \frac{\lambda^2}{s(s+2\lambda)} = \frac{\lambda}{2s} - \frac{1}{4} \cdot \frac{2\lambda}{s+2\lambda},$$

which implies

$$M(t) = \frac{\lambda t}{2} - \frac{1}{4}(1 - e^{-2\lambda t}).$$

The second moment of $N(t)$ about the origin is defined by

$$\begin{aligned}
&E[N(t)^2] \\
&= \sum_{m=1}^{\infty} m^2 P\{N(t) = m\} \\
&= \sum_{m=1}^{\infty} \left[2 \cdot \frac{m(m+1)}{2} - m\right] P\{N(t) = m\} \\
&= 2 \sum_{m=1}^{\infty} \sum_{n=1}^{m} nP\{N(t) = m\} - \sum_{m=1}^{\infty} mP\{N(t) = m\} \\
&= 2 \sum_{n=1}^{\infty} n \sum_{m=n}^{\infty} P\{N(t) = m\} - M(t) \\
&= 2 \sum_{n=1}^{\infty} nP\{N(t) \geq n\} - M(t) \\
&= 2 \sum_{n=1}^{\infty} nP\{S_n \leq t\} - M(t). \qquad (4.2.16)
\end{aligned}$$

Applying the Laplace-Stieltjes transforms to both sides above, we have

$$\int_0^{\infty} e^{-st} dE[N(t)^2]$$

$$= 2\sum_{n=1}^{\infty} n[F^*(s)]^n - M^*(s)$$

$$= 2\left\{\frac{F^*(s)}{1-F^*(s)}\right\}^2 + \frac{F^*(s)}{1-F^*(s)}. \qquad (4.2.17)$$

Inverting the above, we have

$$E[N(t)^2] = 2M * M(t) + M(t), \qquad (4.2.18)$$

and

$$Var(N(t)) = 2M * M(t) + M(t) - [M(t)]^2. \qquad (4.2.19)$$

Example 4.2.6 (*Continuation of Examples 4.2.1 and 4.2.4*) The Laplace-Stieltjes transform of $F(t) = 1 - e^{-\lambda t}$ is given by

$$F^*(s) = \frac{\lambda}{s+\lambda}$$

and

$$\int_0^{\infty} e^{-st} dE[N(t)^2] = 2\left\{\frac{\frac{\lambda}{s+\lambda}}{1-\frac{\lambda}{s+\lambda}}\right\}^2 + \frac{\frac{\lambda}{s+\lambda}}{1-\frac{\lambda}{s+\lambda}}$$

$$= \frac{2\lambda^2}{s^2} + \frac{\lambda}{s},$$

which implies

$$E[N(t)^2] = \lambda^2 t^2 + \lambda t,$$

(see Appendix A), and

$$Var(N(t)) = \lambda t,$$

as shown in Section 3.2 for the Poisson process.

4.3 Limit Theorems

In this section we introduce a few limit theorems for a renewal process. For instance, the so-called *elementary renewal theorem* is well-known. The *key renewal theorem* is also famous and has many applications in practice. First, we discuss the asymptotic behaviors of the renewal function $M(t)$ and variance $Var(N(t))$. Secondly, we give a series of theorems in renewal theory.

Let μ , σ^2, and μ_3 be the mean, variance, and third moment about the origin of the interarrival time distribution $F(t)$, respectively. Expanding $F^*(s)$ with respect to s and noting that the second moment about the origin is $\sigma^2 + \mu^2$, we have

$$F^*(s) = 1 - \mu s + \frac{1}{2}(\sigma^2 + \mu^2)s^2 - \frac{1}{3!}\mu_3 s^3 + o(s^3). \tag{4.3.1}$$

Substituting Eq.(4.3.1) into Eqs.(4.2.15) and (4.2.17), and inverting them, we have

$$M(t) = \frac{t}{\mu} + \left(\frac{\sigma^2}{2\mu^2} - \frac{1}{2}\right) + o(1), \tag{4.3.2}$$

and

$$Var(N(t)) = \frac{\sigma^2 t}{\mu^3} + \left(\frac{1}{12} + \frac{5\sigma^4}{4\mu^4} - \frac{2\mu_3}{3\mu^3}\right) + o(1), \tag{4.3.3}$$

which are asymptotic forms and the bias terms vanish as $t \to \infty$.

Example 4.3.1 For a Poisson process with parameter $\lambda > 0$, we have $M(t) = \lambda t$ and $Var(N(t)) = \lambda t$. That is, the asymptotic forms in Eqs.(4.3.2) and (4.3.3) are justified by eliminating the bias terms. However, if we assume non-exponential interarrival time distributions, the asymptotic forms with the bias terms hold.

We are interested in the asymptotic behavior of $M(t)/t$ as $t \to \infty$. First, we show the following theorem:

Theorem 4.3.1 For a renewal process, with probability one,

$$\frac{N(t)}{t} \longrightarrow \frac{1}{\mu} \tag{4.3.4}$$

as $t \to \infty$, where $\mu = \mathrm{E}[X_m]$ is the mean interarrival time.

Proof Referring to Fig. 3.4.3 for a counting process, we have

$$S_{N(t)} \leq t \leq S_{N(t)+1}, \tag{4.3.5}$$

which implies

$$\frac{S_{N(t)}}{N(t)} \leq \frac{t}{N(t)} \leq \frac{S_{N(t)+1}}{N(t)+1} \cdot \frac{N(t)+1}{N(t)}. \tag{4.3.6}$$

Noting that

$$\frac{S_{N(t)}}{N(t)} = \frac{X_1 + X_2 + \cdots + X_{N(t)}}{N(t)} \tag{4.3.7}$$

and applying the strong law of large numbers in Theorem 2.6.3, we have with probability 1,

$$\frac{t}{N(t)} \longrightarrow \mu \tag{4.3.8}$$

as $t \to \infty$, which proves the theorem.

We are now in a position to show the following most important theorem:

Theorem 4.3.2 (*Elementary Renewal Theorem*) For a renewal process,

$$\frac{M(t)}{t} \longrightarrow \frac{1}{\mu} \tag{4.3.9}$$

as $t \to \infty$, where $\mu = \mathrm{E}[X_m]$ is the mean interarrival time.

Proof As shown in Eq.(4.3.2), we have

$$\frac{M(t)}{t} \longrightarrow \frac{1}{\mu} \tag{4.3.10}$$

as $t \to \infty$, which proves the theorem. However, this proof is a trick since we used the result established in Eq.(4.3.2). To derive Eq.(4.3.2), we have to make recourse to the Tauberian theorem in Appendix A.

We can also show the following theorem without proof (ref. Eq.(4.3.3)):

Theorem 4.3.3 For a renewal process,

$$\frac{Var(N(t))}{t} \longrightarrow \frac{\sigma^2}{\mu^3} \tag{4.3.11}$$

as $t \to \infty$, where μ and σ^2 are the finite mean and variance, respectively, of the interarrival time X_m.

Using the asymptotic forms in Eqs.(4.3.2) and (4.3.3), we have the following theorem which corresponds to the central limit theorem for the renewal process.

Theorem 4.3.4 Let μ and σ^2 be the finite mean and variance, respectively, of the interarrival time distribution $F(t)$ for a renewal process $\{N(t),\ t \geq 0\}$. Then

$$P\left\{\frac{N(t)-t/\mu}{\sqrt{\frac{\sigma^2 t}{\mu^3}}} \leq y\right\} \longrightarrow \Phi(y) = \frac{1}{\sqrt{2\pi}}\int_{-\infty}^{y} e^{-\frac{x^2}{2}}dx \tag{4.3.12}$$

as $t \to \infty$. That is, $N(t)$ is asymptotically normal distributed with mean t/μ and variance $\sigma^2 t/\mu^3$.

Proof Referring to Fig. 3.3.2 for a counting process, we have

$$P\{N(t) < n\} = P\{t < S_n\} \tag{4.3.13}$$

for any $t \geq 0$ and $n \geq 0$. Let n be an integer such that

$$n = \left[\frac{t}{\mu} + y\sqrt{\frac{\sigma^2 t}{\mu^3}}\right] \tag{4.3.14}$$

where the brackets $[\cdot]$ denote an integral part. Then we have

$$P\{N(t) < n\} = P\left\{\frac{N(t)-t/\mu}{\sqrt{\sigma^2 t/\mu^3}} < \frac{n-t/\mu}{\sqrt{\sigma^2 t/\mu^3}}\right\} \tag{4.3.15}$$

which implies

$$\lim_{t\to\infty} P\{N(t) < n\} = \lim_{t\to\infty} P\left\{\frac{N(t)-t/\mu}{\sqrt{\sigma^2 t/\mu^3}} < y\right\}. \tag{4.3.16}$$

On the other hand, we have

$$P\{t < S_n\} = P\left\{\frac{t-n\mu}{\sqrt{n\sigma^2}} < \frac{S_n - n\mu}{\sqrt{n\sigma^2}}\right\}. \tag{4.3.17}$$

Noting that

$$\lim_{t\to\infty} \frac{t - n\mu}{\sqrt{n\sigma^2}} = -y, \tag{4.3.18}$$

we have

$$\lim_{t\to\infty} P\{t < S_n\} = \lim_{n\to\infty} P\left\{-y < \frac{S_n - n\mu}{\sqrt{n\sigma^2}}\right\} \tag{4.3.19}$$

$$= \frac{1}{\sqrt{2\pi}} \int_{-y}^{\infty} e^{-\frac{x^2}{2}} dx = \frac{1}{\sqrt{2\pi}} \int_{-\infty}^{y} e^{-\frac{x^2}{2}} dx = \Phi(y), \tag{4.3.20}$$

which proves

$$\lim_{t\to\infty} P\left\{\frac{N(t) - t/\mu}{\sqrt{\sigma^2 t/\mu^3}} \le y\right\} = \Phi(y). \tag{4.3.21}$$

A non-negative random variable X (or a distribution $F(t)$) is said to be *lattice* or *arithmetic* if there exists $\delta > 0$ such that $\sum_{n=0}^{\infty} P\{X = n\delta\} = 1$, where δ is called the period of X. The following theorem is well-known as Blackwell's theorem.

Theorem 4.3.5 (*Blackwell's Theorem*)

(i) If $F(t)$ is *not lattice*, then for all $\tau \ge 0$

$$M(t + \tau) - M(t) \longrightarrow \frac{\tau}{\mu} \tag{4.3.22}$$

as $t \to \infty$.

(ii) If $F(t)$ is *lattice* with period δ, then

$$M(n\delta) - M((n-1)\delta) \longrightarrow \frac{\delta}{\mu} \tag{4.3.23}$$

as $n \to \infty$.

Since the proof needs elaborate analysis, we will forego proving the theorem. However, it is quite easy to understand Eq.(4.3.22). It implies that the mean number of renewals within the time interval $(t,\ t + \tau]$ tends to $\frac{\tau}{\mu}$ as $t \to \infty$. In other words, as $t \to \infty$, the mean number of renewals is proportional to its time duration, where the proportional constant is $1/\mu$, the reciprocal of the mean interarrival time.

Let $h(t)$ be an arbitrary and bounded function defined between $[0, \infty)$. Let $\underline{m_k}(a)$ and $\overline{m_k}(a)$ be the infimum and supremum over the interval $(k-1)a \leq t \leq ka$ for any $a \geq 0$. Then $h(t)$ is said to be *directly Riemann integrable* if $\sum_{k=1}^{\infty} \underline{m_k}(a)$ and $\sum_{k=1}^{\infty} \overline{m_k}(a)$ are finite for all $a \geq 0$ and converge toward the same value as $a \to 0$. We will simply cite a *sufficient* condition for $h(t)$ being directly Riemann integrable:

(i) $h(t) \geq 0$ for all $t \geq 0$.

(ii) $h(t)$ is non-increasing.

(iii) $\int_0^{\infty} h(t)dt < \infty$.

We are now in a position to show the key renewal theorem (without proof) which will be of great use in applied probability models.

Theorem 4.3.6 (*The Key Renewal Theorem*) If $F(t)$ is not lattice and $h(t)$ is directly Riemann integrable, then

$$\lim_{t\to\infty} \int_0^t h(t-x)dM(x) = \frac{1}{\mu}\int_0^{\infty} h(t)dt, \tag{4.3.24}$$

where $M(t)$ is the renewal function.

To understand the key renewal theorem more clearly, we recall Blackwell's theorem: If $F(t)$ is not lattice, we have

$$\lim_{t\to\infty} \frac{M(t+\tau)-M(t)}{\tau} = \frac{1}{\mu}. \tag{4.3.25}$$

Assuming $\tau \to 0$ in Eq.(4.3.25), we have

$$\lim_{\tau\to 0}\lim_{t\to\infty} \frac{M(t+\tau)-M(t)}{\tau} = \lim_{t\to\infty} \frac{dM(t)}{dt} = \frac{1}{\mu}, \tag{4.3.26}$$

where interchanging the limits above can be justified. The function $m(t) = dM(t)/dt$ is called the *renewal density* and represents the mean number of renewals per unit time at time t. Referring to Eq.(4.3.26), we can now better understand the key renewal theorem in Eq.(4.3.24).

We shall show how to apply the key renewal theorem in practice. First, we can generalize the renewal equation to

$$g(t) = h(t) + \int_0^t g(t-x)dF(x) \qquad (t \geq 0), \tag{4.3.27}$$

where $h(t)$ and $F(t)$ are known and $g(t)$ is unknown in the integral equation. The integral equation (4.3.27) is called the *renewal-type equation* and is given in terms of the renewal function $M(t)$:

$$g(t) = h(t) + \int_0^t h(t-x)dM(x), \tag{4.3.28}$$

where $M(t) = \sum_{n=1}^{\infty} F^{(n)}(t)$ is the renewal function. It is quite easy to derive Eq.(4.3.28) from Eq.(4.3.27). Let $g^*(s)$ and $h^*(s)$ denote the Laplace-Stieltjes transforms of $g(t)$ and $h(t)$, respectively. Taking the Laplace-Stieltjes transforms on both sides of Eq.(4.3.27) implies

$$\begin{aligned} g^*(s) &= \frac{h^*(s)}{1-F^*(s)} \\ &= h^*(s) + h^*(s) \cdot \frac{F^*(s)}{1-F^*(s)} \\ &= h^*(s) + h^*(s) \cdot M^*(s). \end{aligned} \tag{4.3.29}$$

Inverting the Laplace-Stieltjes transforms on both sides of Eq.(4.3.29) implies Eq.(4.3.28). Once the renewal-type equation is given and the suitable conditions are satisfied (e.g., $h(t)$ is directly Riemann integrable), we can apply the key renewal theorem to Eq.(4.3.28), which implies the asymptotic behavior of $g(t)$ as $t \to \infty$. Moreover, we can also show how to apply the renewal-type equation in practice.

A direct application of the results is found in the residual life distribution $P\{\gamma_t \le x\}$ (see Fig. 3.4.3). On the condition that the first renewal takes place at $X_1 = y$, we have

$$P\{\gamma_t > x \mid X_1 = y\} = \begin{cases} P\{\gamma_{t-y} > x\} & (y \le t) \\ 0 & (t < y \le t+x) \\ 1 & (y > t+x). \end{cases} \tag{4.3.30}$$

Consult Fig. 4.3.1 for understanding Eq.(4.3.30).

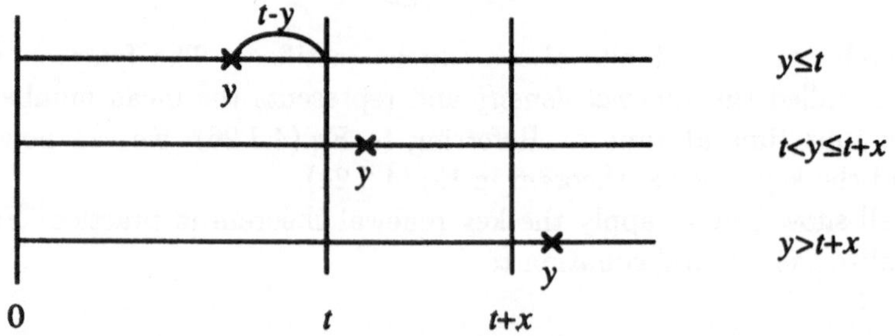

Fig. 4.3.1 The possible cases of $P\{\gamma_t > x \mid X_1 = y\}$.

The residual life distribution $P\{\gamma_t > x\}$ is given by

$$P\{\gamma_t > x\} = \int_0^\infty P\{\gamma_t > x \mid X_1 = y\} dF(y). \tag{4.3.31}$$

Substituting Eq.(4.3.30) into Eq.(4.3.31) implies

$$\begin{aligned} P\{\gamma_t > x\} &= \int_0^t P\{\gamma_{t-y} > x\} dF(y) + \int_t^{t+x} 0 \cdot dF(y) + \int_{t+x}^\infty 1 \cdot dF(y) \\ &= 1 - F(t+x) + \int_0^t P\{\gamma_{t-y} > x\} dF(y), \end{aligned} \tag{4.3.32}$$

which is a renewal-type equation. Let us define $h(t) = 1 - F(t+x)$. Then $h(t)$ is directly Riemann integrable. Using $h(t)$ in Eq.(4.3.32), we can obtain

$$P\{\gamma_t > x\} = h(t) + \int_0^t h(t-y) dM(y) \tag{4.3.33}$$

(see Eq.(4.3.28)). Applying the key renewal theorem in Theorem 4.3.6 and noting that $\lim_{t\to\infty} h(t) = 0$, we have

$$\begin{aligned} \lim_{t\to\infty} P\{\gamma_t > x\} &= \frac{1}{\mu} \int_0^\infty [1 - F(t+x)] dt \\ &= \frac{1}{\mu} \int_x^\infty [1 - F(y)] dy. \end{aligned} \tag{4.3.34}$$

It is clear that

$$P\{\gamma_t \le x\} = 1 - P\{\gamma_t > x\}, \tag{4.3.35}$$

and

$$\lim_{t\to\infty} P\{\gamma_t \le x\} = 1 - \lim_{t\to\infty} P\{\gamma_t > x\}. \tag{4.3.36}$$

Summarizing the above facts, we have the following (see Problem 4.7):

Theorem 4.3.7 The excess or residual life distribution is

$$P\{\gamma_t \le x\} = F(t+x) - \int_0^t [1 - F(t+x-y)] dM(y), \tag{4.3.37}$$

and its asymptotic distribution is

$$\lim_{t\to\infty} P\{\gamma_t \le x\} = \frac{1}{\mu}\int_0^x [1 - F(y)]dy, \tag{4.3.38}$$

if $F(t)$ is *not lattice.*

Let us consider the current life or age distribution $P\{\delta_t \le x\}$. Noting that

$$\delta_t > x \iff \text{no renewals in } [t - x,\ t], \tag{4.3.39}$$

and

$$\gamma_{t-x} > x \iff \text{no renewals in } [t - x,\ t], \tag{4.3.40}$$

we have

$$P\{\delta_t > x\} = P\{\gamma_{t-x} > x\}, \tag{4.3.41}$$

Applying the results in Eqs.(4.3.37) and (4.3.38), we have the following theorem.

Theorem 4.3.8 The current life or age distribution is

$$P\{\delta_t \le x\} = \begin{cases} F(t) - \int_0^{t-x} [1 - F(t - y)]dM(y) & (x \le t) \\ 1 & (x > t), \end{cases} \tag{4.3.42}$$

and its asymptotic distribution is

$$\lim_{t\to\infty} P\{\delta_t \le x\} = \frac{1}{\mu}\int_0^x [1 - F(y)]dy, \tag{4.3.43}$$

if $F(t)$ is *not lattice.*

In Eq.(4.3.42), it is obvious that the probability that $\delta_t \le x$ for $x > t$ is unity, since it is always true that $\delta_t \le x$ for $x > t$.

It is quite interesting that both the asymptotic current life and excess life distributions are given by the same distribution

$$F_e(t) = \frac{1}{\mu}\int_0^t [1 - F(y)]dy, \tag{4.3.44}$$

which is called the *equilibrium distribution* of $F(t)$.

Example 4.3.2 A person wants to take a bus at a bus stop. The arrival law of buses at a bus stop follows a renewal process $\{N(t),\ t \ge 0\}$, where the interarrival time distribution is $F(t)$ with finite mean $1/\mu$ and variance σ^2. How long does he wait for a bus to come if he arrives at the bus stop at random? The waiting time distribution is given by the asymptotic residual life distribution since he arrives at the bus stop at random, which means that the original starting time of the renewal process is far removed from his arrival time. Thus his waiting time distribution is given by

$$W(x) = \lim_{t\to\infty} P\{\gamma_t \leq x\} = \frac{1}{\mu}\int_0^x [1 - F(y)]dy,$$

and the mean waiting time is given by

$$\int_0^\infty x dW(x) = \frac{1}{\mu}\int_0^\infty x[1 - F(x)]dx = \frac{\mu^2 + \sigma^2}{2\mu} = \frac{\mu}{2} + \frac{\sigma^2}{2\mu}.$$

The intuitive answer would be $\mu/2$, half of the mean interarrival time. However, the actual answer is $\frac{\mu}{2} + \frac{\sigma^2}{2\mu}$, half of the mean interarrival time plus the term of variance. Of course, if the buses arrive regularly, i.e., each bus arrives just at μ, the intuitive answer is correct. However, in general, we have to consider the term of variance.

4.4 Delayed and Stationary Renewal Processes

In the preceding sections we have assumed that all interarrival distributions are identical. However, we can generalize that only the first interarrival time distribution, say $G(t)$, is different from successive interarrival time distributions $F(t)$.

Definition 4.4.1 A counting process $\{N_D(t),\ t \geq 0\}$ is called a *delayed renewal process* if the interarrival times are independent and the first interarrival time is distributed acorrding to $G(t)$ and the successive interarrival times are distributed identically according to $F(t)$.

The renewal process in the preceding sections has been discussed by assuming independent and identical distribution $F(t)$. It is easy to show the following results for a delayed renewal process:

$$P\{N_D(t) = m\} = \begin{cases} 1 - G(t) & (m = 0) \\ G * F^{(m-1)}(t) - G * F^{(m)}(t) & (m = 1,\ 2,\ \cdots), \end{cases} \tag{4.4.1}$$

and a renewal function

$$M_D(t) = E[N_D(t)] = \sum_{m=0}^{\infty} G * F^{(m)}(t), \tag{4.4.2}$$

where $F^{(m)}(t)$ is defined in Eq.(4.2.4). The renewal equation is given by

$$M_D(t) = G(t) + \int_0^t M_D(t-x)dF(x) \tag{4.4.3}$$

which is a renewal-type equation introduced in Eq.(4.3.27). Of course, the Laplace-Stieltjes transform of $M_D(t)$ is given by

$$M_D^*(s) = \frac{G^*(s)}{1-F^*(s)}, \tag{4.4.4}$$

where $M_D^*(s)$ and $G^*(s)$ are the Laplace-Stieltjes transforms of $M_D(t)$ and $G(t)$, respectively.

The following theorem can be easily verified from the preceding results of the renewal process. In particular, the asymptotic results are just the same since the first interarrival time distribution $G(t)$ has no effects on the asymptotic results:

Theorem 4.4.1 For a delayed renewal process $\{N_D(t),\ t \geq 0\}$:

(i) With probability 1,

$$\frac{N_D(t)}{t} \longrightarrow \frac{1}{\mu} \tag{4.4.5}$$

as $t \to \infty$, where μ is the mean interarrival time of $F(t)$.

(ii) (*Elementary Renewal Theorem*)

$$\frac{M_D(t)}{t} \longrightarrow \frac{1}{\mu} \tag{4.4.6}$$

as $t \to \infty$.

(iii) (*Blackwell's Theorem*) If $F(t)$ is *not lattice*, then for all $\tau \geq 0$

$$M_D(t+\tau) - M_D(t) \longrightarrow \frac{\tau}{\mu} \tag{4.4.7}$$

as $t \to \infty$, and if $G(t)$ and $F(t)$ are *lattice* with same period δ,

$$M_D(n\delta) - M_D((n-1)\delta) \longrightarrow \frac{\delta}{\mu} \tag{4.4.8}$$

as $n \to \infty$.

(iv)

$$P\{\gamma_t \le x\} = G(t+x) - \int_0^t [1 - F(t+x-y)]dM_D(y), \tag{4.4.9}$$

(v) If $F(t)$ is *not lattice,*

$$\lim_{t\to\infty} P\{\gamma_t \le x\} = \frac{1}{\mu}\int_0^x [1 - F(y)]dy. \tag{4.4.10}$$

(vi)

$$P\{\delta_t \le x\} = \begin{cases} G(t) - \int_0^{t-x} [1 - F(t-y)]dM_D(y) & (x \le t) \\ 1 & (x > t). \end{cases} \tag{4.4.11}$$

(vii) If $F(t)$ is *not lattice,*

$$\lim_{t\to\infty} P\{\delta_t \le x\} = \frac{1}{\mu}\int_0^x [1 - F(y)]dy. \tag{4.4.12}$$

As a special case of a delayed renewal process, we assume that the first interarrival time distribution is given by

$$G(t) = F_e(t) = \frac{1}{\mu}\int_0^t [1 - F(y)]dy. \tag{4.4.13}$$

Such a renewal process can be obtained from the underlying renewal process if we assume the starting point (time 0) sufficiently far from the time origin of the underlying renewal process. Then the first interarrival time distribution is given by the equilibrium distribution in Eq.(4.3.44).

Definition 4.4.2 A counting process $\{N_S(t),\ t \ge 0\}$ is called a *stationary renewal process* if the interarrival times are independent and the first interarrival time is distributed with

$$F_e(t) = \frac{1}{\mu}\int_0^t [1 - F(y)]dy, \tag{4.4.14}$$

and the successive interarrival times are distributed identically with $F(t)$, where $\mu = \int_0^\infty t dF(t)$ is the mean interarrival time.

The Laplace-Stieltjes transform of $F_e(t)$ is given by

$$F_e^*(s) = \frac{1 - F^*(s)}{\mu s}, \tag{4.4.15}$$

and the Laplace-Stieltjes transform of the renewal function $M_S(t) = E[N_S(t)]$ is given by

$$M_S^*(s) = \frac{F_e^*(s)}{1 - F^*(s)} = \frac{1 - F^*(s)}{\mu s} \cdot \frac{1}{1 - F^*(s)} = \frac{1}{\mu s}, \tag{4.4.16}$$

which implies

$$M_S(t) = \frac{t}{\mu}. \tag{4.4.17}$$

It is surprising that the renewal function is just the same as the Poisson process with parameter $1/\mu$.

Substituting $G(t) = F_e(t)$ and $M_D(t) = M_S(t) = t/\mu$ into Eq.(4.4.9), we have for all t,

$$\begin{aligned} P\{\gamma_t \le x\} &= \frac{1}{\mu}\int_0^{t+x} [1 - F(y)]dy - \int_0^t [1 - F(t + x - y)]\frac{dy}{\mu} \\ &= \frac{1}{\mu}\left\{\int_0^{t+x} [1 - F(y)]dy - \int_x^{t+x} [1 - F(y)]dy\right\} \\ &= \frac{1}{\mu}\int_0^x [1 - F(y)]dy, \end{aligned} \tag{4.4.18}$$

which is just the same as its asymptotic distribution in Eq.(4.4.10). Similar to the discussion above, we have for all t,

$$P\{\delta_t \le x\} = \begin{cases} \frac{1}{\mu}\int_0^x [1 - F(y)]dy & (x \le t) \\ 1 & (x > t). \end{cases} \tag{4.4.19}$$

Summarizing the results above, we have the following:

Theorem 4.4.2 For a stationary renewal process $\{N_S(t),\ t \ge 0\}$:

(i)

$$M_S(t) = E[N_S(t)] = \frac{t}{\mu}, \tag{4.4.20}$$

(ii)

$$P\{\gamma_t \le x\} = \frac{1}{\mu}\int_0^x [1 - F(y)]dy, \tag{4.4.21}$$

for all $t \ge 0$.

(iii)

$$P\{\delta_t \le x\} = \begin{cases} \frac{1}{\mu}\int_0^x [1 - F(y)]dy & (x \le t) \\ 1 & (x > t), \end{cases} \tag{4.4.22}$$

for all $t \ge 0$.

(iv) For all $t,\ \tau \ge 0$.

$$P\{N_S(t + \tau) - N_S(\tau) = m\} = P\{N_S(t) = m\}, \tag{4.4.23}$$

i.e., the process $\{N_S(t),\ t \ge 0\}$ has *stationary independent increments.*

4.5 Problems 4

4.1 Show that the following formula for interchanging the orders of summation is correct:

$$\sum_{m=1}^{\infty}\sum_{n=1}^{m} a_{mn} = \sum_{n=1}^{\infty}\sum_{m=n}^{\infty} a_{mn}.$$

4.2 (*Examples 4.2.2 and 4.2.3*) The renewal function $M(t)$ was given in Example 4.2.3 for $X_m \sim GAM(\lambda,\ k)$ $(m = 1,\ 2,\ \cdots)$ in general. Show that

$$M(t) = \lambda t \qquad (k = 1),$$

$$M(t) = \frac{\lambda t}{2} - \frac{1}{4}(1 - e^{-2\lambda t}) \qquad (k = 2),$$

are special cases of the general formula.

4.3 (*Example 4.2.2*) Derive the variance $Var(N(t))$ in Eq.(4.2.19) for a renewal process with $X_m \sim GAM(\lambda,\ 2)$ $(m = 1,\ 2,\ \cdots)$.

4.4 Consider a renewal process $\{N(m),\ m = 1,\ 2,\ \cdots\}$ with (discrete) geometric interarrival distribution having parameter p, i.e., $X_m \sim GEO(p)$ $(m = 1,\ 2,\ \cdots)$.

(i) Derive the Laplace-Stieltjes transform $F^*(s)$.

(ii) Derive the Laplace-Stieltjes transform $M^*(s)$ in Eq.(4.2.13) and invert it.

4.5 (*Continuation*) Consider a renewal process $\{N(m),\ m = 2,\ 3,\ \cdots\}$ with (discrete) negative binomial distribution of order 2, i.e., $X_m \sim NB(p,\ 2)$ $(m = 1,\ 2,\ \cdots)$. Derive $M(m)$ $(m = 2,\ 3,\ \cdots)$ by using the routine described in Problem 4.4.

4.6 Let $m(t) = dM(t)/dt$ denote the renewal density for a renewal process with interarrival time distribution $F(t)$. If there exists the density $f(t) = dF(t)/dt$, verify that $m(t)$ satisfies

$$m(t) = f(t) + \int_0^t m(t-x)f(x)dx,$$

which is also called the *renewal equation*.

4.7 Derive Eqs.(4.3.37) and (4.3.38) from Eqs.(4.3.33) and (4.3.34), respectively.

4.8 Derive Eqs.(4.3.42) and (4.3.43) from Eq.(4.3.41) and Theorem 4.3.7.

4.9 Let μ and σ^2 be a finite mean and variance for the interarrival time distribution $F(t)$ in a renewal process.

(i) Derive the renewal-type equation for $g(t) = M(t) - t/\mu + 1$ and obtain $h(t)$ in Eq.(4.3.27).

(ii) Verify

$$\lim_{t\to\infty}\left[M(t) - \frac{t}{\mu} + 1\right] = \frac{\sigma^2 + \mu^2}{2\mu^2}$$

by applying the key renewal theorem.

Chapter 5

Discrete-Time Markov Chains

5.1 Introduction

In the preceding two chapters we have discussed two continuous-time stochastic processes, i.e., the Poisson process and the renewal process. In this and in following chapters we shall discuss Markov chains. Recall that we are considering stochastic processes with discrete-state space throughout this book. Let us define a discrete-time Markov chain:

Definition 5.1.1 Let $\{X(n), n = 0,\ 1,\ 2,\ \cdots\}$ be a discrete-time stochastic process with state space $i = 0,\ 1,\ 2,\ \cdots$, unless otherwise specified. If

$$P\{X(n+1) = j \mid X(0) = i_0,\ X(1) = i_1,\ \cdots,\ X(n-1) = i_{n-1},\ X(n) = i\}$$
$$= P\{X(n+1) = j \mid X(n) = i\} = p_{ij} \tag{5.1.1}$$

for all $i_0,\ i_1,\ \cdots,\ i_{n-1},\ i,\ j$ and n, then the process is called a *discrete-time Markov chain* and p_{ij} is called a (*stationary*) *transition probability*.

Note that $X(n) = i$ denotes that the process is in state i ($i = 0,\ 1,\ 2,\ \cdots$) at time n ($n = 0,\ 1,\ 2,\ \cdots$). The conditional probability in Eq.(5.1.1), which describes the whole history, is independent of the past history at time $0,\ 1,\ 2,\ \cdots,\ n-1$ and depends on the present state $X(n) = i$. That is, the conditional probability of moving to the "future" state depends only on the "present" state and is independent of "past" history. Such a property is called the *Markov property*.

The transition probability p_{ij} in Eq.(5.1.1) depends on the present time n in general. However, we assume that the transition probability is *stationary*, i.e., independent of the present time n. That is, such a process is precisely called a *Markov chain with stationary transition probabilities.* Throughout this chapter we discuss such a process.

A matrix $\mathbf{P} = [p_{ij}]$ is called a *transition probability matrix*, where

$$p_{ij} \geq 0, \qquad \sum_{j=0}^{\infty} p_{ij} = 1 \qquad (i, j = 0,\ 1,\ 2,\ \cdots). \tag{5.1.2}$$

Two simple examples of a Markov chain are illustrated.

Example 5.1.1 Consider a two-state Markov chain with the following transition probability matrix:

$$\mathbf{P} = \begin{bmatrix} p_{00} & p_{01} \\ p_{10} & p_{11} \end{bmatrix} = \begin{bmatrix} 0 & 1 \\ 1 & 0 \end{bmatrix}.$$

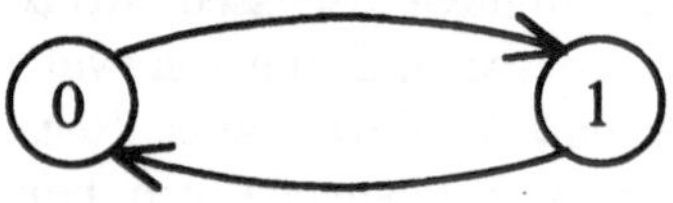

Fig. 5.1.1 A state transition diagram of Example 5.1.1.

That is, at any given time the process moves to another state with the probability of 1. Such behavior is illustrated by using the so-called "state transition diagram" in Fig. 5.1.1, where the number *circled* denotes a state and the number appearing on an arc is the transition probability. We can understand the behavior more clearly from the state transition diagram.

Example 5.1.2 Consider again a two-state Markov chain. In general, the transition probability matrix is given by

$$\mathbf{P} = \begin{bmatrix} p_{00} & p_{01} \\ p_{10} & p_{11} \end{bmatrix} = \begin{bmatrix} 1-a & a \\ b & 1-b \end{bmatrix},$$

where $0 \leq a \leq 1$, $0 \leq b \leq 1$, and $| 1 - a - b | < 1$. A state transition diagram is shown in Fig. 5.1.2. Relaxing a condition that $| 1 - a - b | < 1$ in this example, we have Example 1.5.1 with $a = b = 1$.

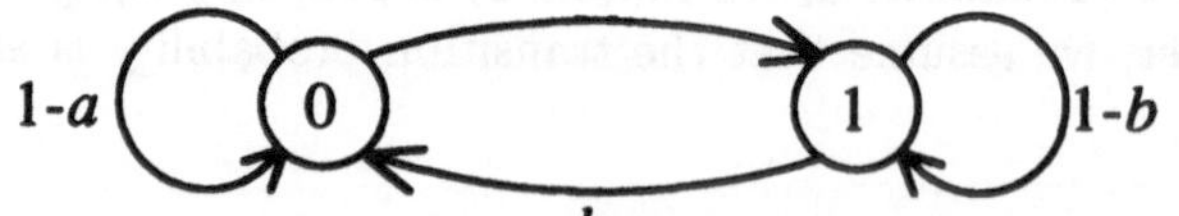

Fig. 5.1.2 A state transition diagram of Example 5.1.2.

5.2 Chapman-Kolmogorov Equation

In general we assume that the number of states in a Markov chain is *denumberable* or *infinitely countable* except in Section 5.5. Of course, we cite several examples of a finite-state Markov chain for illustration. However, the results are valid for a denumberable Markov chain.

Recall that the Markov chain is specified by the transition probability matrix $\mathbf{P} = [p_{ij}]$, where the law of the total probability implies

$$p_{ij} \geq 0, \qquad \sum_{j=0}^{\infty} p_{ij} = 1 \qquad (i,\ j,\ = 0,\ 1,\ 2,\ \cdots). \tag{5.2.1}$$

The probabilistic behavior of a Markov chain $\{X(n),\ n = 0,\ 1,\ 2,\ \cdots\}$ can be completely described by the initial probability and the transition probabilities as follows:

$$\begin{aligned} & P\{X(0) = i_0,\ X(1) = i_1,\ \cdots,\ X(n) = i_n\} \\ &= P\{X(n) = i_n \mid X(0) = i_0,\ X(1) = i_1,\ \cdots,\ X(n-1) = i_{n-1}\} \\ &\quad \cdot P\{X(0) = i_0,\ X(1) = i_1,\ \cdots,\ X(n-1) = i_{n-1}\} \\ &= p_{i_{n-1}i_n} P\{X(0) = i_0,\ X(1) = i_1,\ \cdots,\ X(n-1) = i_{n-1}\} \\ &= \cdots \\ &= p_{i_{n-1}i_n} p_{i_{n-2}i_{n-1}} \cdots p_{i_0 i_1} P\{X(0) = i_0\} \end{aligned} \tag{5.2.2}$$

where $P\{X(0) = i_0\}$ is the *initial probability*. Let $\pi(0)$ denote the *initial distribution*

$$\pi(0) = [\pi_0(0),\ \pi_1(0),\ \cdots], \tag{5.2.3}$$

where

$$\pi_j(0) = P\{X(0) = j\} \geq 0 \qquad (j = 0,\ 1,\ 2,\ \cdots) \tag{5.2.4}$$

is the initial probability such that

$$\sum_{j=0}^{\infty} \pi_j(0) = 1. \tag{5.2.5}$$

As will be shown later, we can calculate all the transition probabilities by specifying the transition probability matrix $\mathbf{P} = [p_{ij}]$ and the initial distribution $\pi(0)$.

Let

$$p_{ij}^n = P\{X(n+m) = j \mid X(m) = i\} \tag{5.2.6}$$

denote the *n-step transition probability* that the process is in state j at time $n+m$ given that it was in state i at time m. Note that the n-step transition probability p_{ij}^n is independent of the present time m and depends only on the time duration n. Our interest is to calculate p_{ij}^n in terms of the transition probability p_{ij}, where we postulate

$$p_{ij}^0 = 0 \qquad (i \neq j), \qquad p_{ii}^0 = 1, \tag{5.2.7}$$

for convenience.

The n-step transition probability p_{ij}^n can be calculated by summing over all the intermediate state k at time r $(0 \le r \le n)$ and moving to state j from state k at the remaining time $n-r$ (see Fig. 5.2.1). That is,

$$p_{ij}^n = \sum_{k=0}^{\infty} p_{ik}^r p_{kj}^{(n-r)}, \tag{5.2.8}$$

which is called the *Chapman-Kolmogorov equation*. In a matrix form, introducing $\mathbf{P}^{(n)} = [p_{ij}^n]$, we have

$$\mathbf{P}^{(n)} = \mathbf{P}^{(r)} \cdot \mathbf{P}^{(n-r)}. \tag{5.2.9}$$

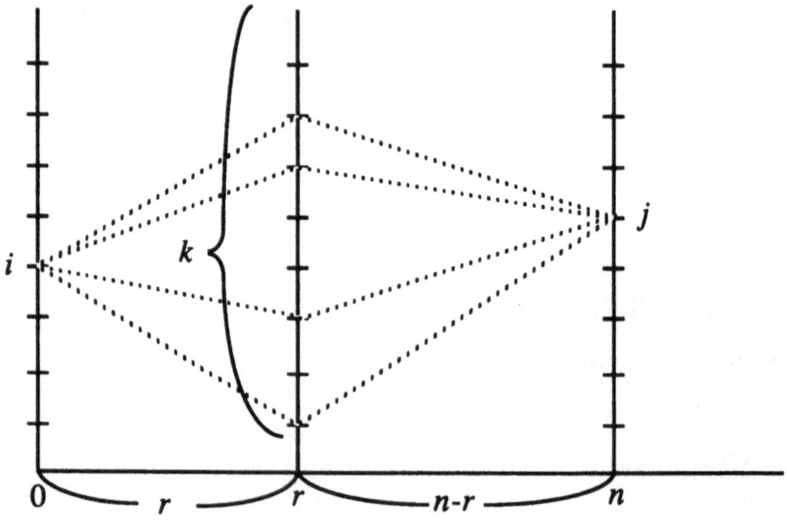

Fig. 5.2.1 An interpretation of Eq.(5.2.8).

Noting that $\mathbf{P}^{(1)} = \mathbf{P} = [p_{ij}]$, we have recursively

$$\mathbf{P}^{(n)} = \mathbf{P}^{(1)} \cdot \mathbf{P}^{(n-1)} = \mathbf{P} \cdot \mathbf{P}^{(n-1)} = \mathbf{P}^2 \cdot \mathbf{P}^{(n-2)} = \cdots = \mathbf{P}^n. \qquad (5.2.10)$$

That is, the n-step transition probability can be calculated by the nth power of matrix $\mathbf{P}$. Rewriting Eq.(5.2.9) by using the above fact, we have

$$\mathbf{P}^n = \mathbf{P}^r \cdot \mathbf{P}^{n-r} \qquad (5.2.11)$$

which is merely a matrix product. The Chapman-Kolmogorov equation plays a central role in developing the theory of the discrete-time Markov chain.

As shown in Eq.(5.2.2), the joint probability can be calculated by specifying the initial probability (i.e., the initial distribution $\pi(0)$) and the transition probability (i.e., the transition probability matrix $\mathbf{P}$). Let

$$\begin{aligned} \pi_j(n) &= P\{X(n) = j\} \\ &= \sum_{i=0}^{\infty} P\{X(n) = j \mid X(0) = i\} P\{X(0) = i\} \\ &= \sum_{i=0}^{\infty} \pi_i(0) p_{ij}^n \qquad (j = 0,\ 1,\ 2,\ \cdots) \end{aligned} \qquad (5.2.12)$$

be the probability that the process is in state j at time n. Let

$$\pi(n) = [\pi_0(n),\ \pi_1(n),\ \cdots] \qquad (5.2.13)$$

be the *n-step distribution* such that

$$\sum_{j=0}^{\infty} \pi_j(n) = 1. \qquad (5.2.14)$$

Then we can calculate the following matrix form:

$$\pi(n) = \pi(0)\mathbf{P}^n. \qquad (5.2.15)$$

Example 5.2.1 (*Example 5.1.2*) Consider a numerical example when $a = 0.2$ and $b = 0.3$. That is,

$$\mathbf{P} = \begin{bmatrix} 0.8 & 0.2 \\ 0.3 & 0.7 \end{bmatrix}.$$

Figure 5.2.2 shows the curves of p_{ij}^n as $n \to \infty$.

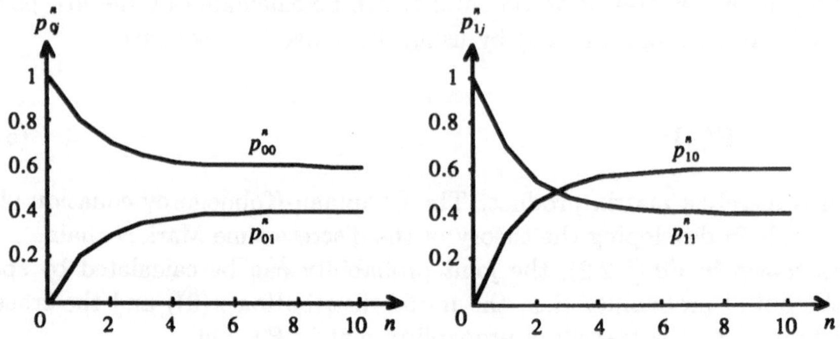

Fig. 5.2.2 Curves of p_{ij}^n as $n \to \infty$ ($a = 0.2$, $b = 0.3$).

Example 5.2.2 (*Example 5.1.2*) Consider another numerical example when $a = 0.6$ and $b = 0.9$. That is,

$$\mathbf{P} = \begin{bmatrix} 0.4 & 0.6 \\ 0.9 & 0.1 \end{bmatrix}.$$

Figure 5.2.3 shows the curves of p_{ij}^n as $n \to \infty$ ($a = 0.6$, $b = 0.9$).

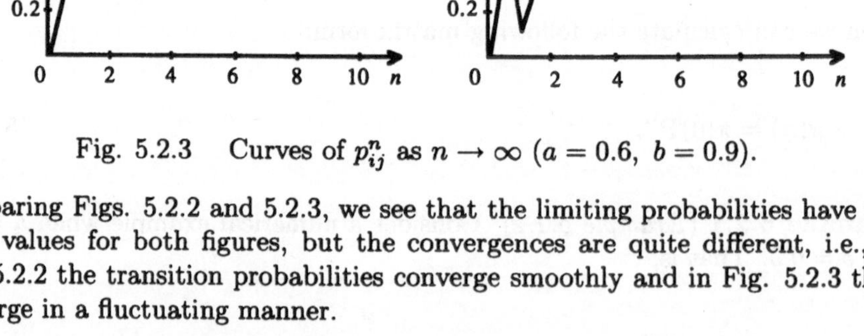

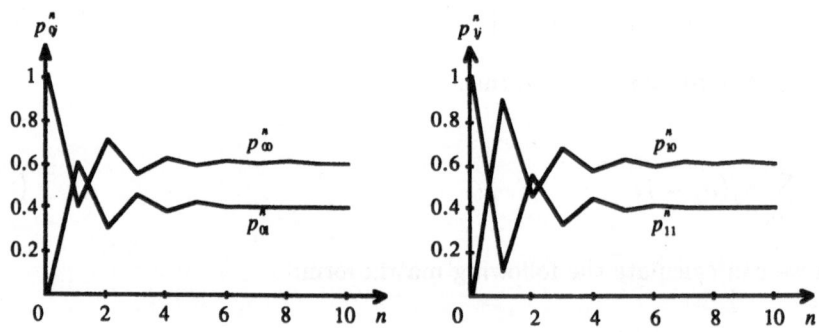

Fig. 5.2.3 Curves of p_{ij}^n as $n \to \infty$ ($a = 0.6$, $b = 0.9$).

Comparing Figs. 5.2.2 and 5.2.3, we see that the limiting probabilities have the same values for both figures, but the convergences are quite different, i.e., in Fig. 5.2.2 the transition probabilities converge smoothly and in Fig. 5.2.3 they converge in a fluctuating manner.

Let us recall Example 5.1.2 in general. We can calculate

$$\mathbf{P}^2 = \begin{bmatrix} 1-a & a \\ b & 1-b \end{bmatrix} \cdot \begin{bmatrix} 1-a & a \\ b & 1-b \end{bmatrix}$$

$$= \begin{bmatrix} (1-a)^2 + ab & (1-a)a + (1-b)a \\ (1-a)b + (1-b)b & ab + (1-b)^2 \end{bmatrix}. \tag{5.2.16}$$

In general, the n-step transition probability matrix is given by

$$\mathbf{P}^n = \frac{1}{a+b}\begin{bmatrix} b & a \\ b & a \end{bmatrix} + \frac{(1-a-b)^n}{a+b}\begin{bmatrix} a & -a \\ -b & b \end{bmatrix}. \tag{5.2.17}$$

Equation (5.2.17) can be verified by the principle of mathematical induction (see Problem 5.6). Applying the results in Eq.(5.2.17), we can understand the behavior in Figs. 5.2.2 and 5.2.3. That is, in Example 5.2.1, we have $1-a-b = 0.5 > 0$, and in Example 5.2.2, we have $1-a-b = -0.5 < 0$, which implies that the transition probabilities fluctuate around and converge toward the limiting probabilities as $n \to \infty$.

Example 5.2.3 Consider a two-state Markov chain with the following transition matrix:

$$\mathbf{P} = \begin{bmatrix} p_{00} & p_{01} \\ p_{10} & p_{11} \end{bmatrix} = \begin{bmatrix} 1-a & a \\ 1-a & a \end{bmatrix},$$

where $0 < a < 1$. Then

$$\mathbf{P}^n = \begin{bmatrix} 1-a & a \\ 1-a & a \end{bmatrix}$$

for any $n > 0$. That is, the transition probability is independent of the present state. Of course,

$$\pi(n) = \pi(0)\mathbf{P}^n = [1-a, \ a].$$

In general, we assume that the transition probability is given by

$$p_{ij} = p_{\cdot j} > 0 \qquad (i, \ j = 0, \ 1, \ \cdots). \tag{5.2.18}$$

That is, the transition probability is positive and independent of the present state i. Such a Markov chain is called a *spatially homogeneous Markov chain* and will appear in Section 5.4.

5.3 State Classification

Two numerical examples in Examples 5.2.1 and 5.2.2 suggest to us that the transition probability p_{ij}^n converges toward the limiting probability π_j, say, as $n \to \infty$, which is independent of the initial state i. Before discussing the limiting behavior of the transition probability, we first discuss the ***state classification*** for a Markov chain.

Consider a Markov chain $\{X(n),\ n = 0,\ 1,\ 2,\ \cdots\}$. If there exists an integer $n \geq 0$ such that $p_{ij}^n > 0$, then state j is *accessible* from state i and denote it $i \to j$. Of course, any state is accessible from itself, i.e., $i \to i$ since $p_{ii}^0 = 1$ in Eq.(5.2.7). If $i \to j$ and $j \to i$, i.e., there exist integers $m \geq 0$ and $n \geq 0$ such that $p_{ij}^m > 0$ and $p_{ji}^n > 0$, then states i and j *communicate*, and denote it $i \leftrightarrow j$. By using a *communication relation*, we can classify all the states of a Markov chain into equivalence classes.

Theorem 5.3.1 Communication is an equivalence relation. That is,

(i) $i \leftrightarrow i$.

(ii) If $i \leftrightarrow j$, then $j \leftrightarrow i$.

(iii) If $i \leftrightarrow j$ and $j \leftrightarrow k$, then $i \leftrightarrow k$.

Proof The first two items are obvious from the definition of communication. To prove **(iii)**, if we assume that there exist integers m and n such that $p_{ij}^m > 0$ and $p_{jk}^n > 0$, then from the Chapman-Kolmogorov equation

$$p_{ik}^{m+n} = \sum_{l=0}^{\infty} p_{il}^m p_{lk}^n \geq p_{ij}^m p_{jk}^n > 0,$$

which implies $i \to k$. Similarly, we can prove $k \to i$, which proves **(iii)**.

By means of a communication relation, all the states in a Markov chain are classified into equivalence classes which are disjoint and exhaustive.

Example 5.3.1 (*Examples 5.1.1 and 5.1.2*) For both examples (except $a = b = 0$), we have $0 \to 1$ and $1 \to 0$, which implies one class $\{0,\ 1\}$. Of course, $a = b = 0$ of Example 5.1.2 implies two classes $\{0\}$ and $\{1\}$.

Example 5.3.2 Consider a Markov chain with the following transition probability matrix:

$$\mathbf{P} = \begin{matrix} 0 \\ 1 \\ 2 \end{matrix} \begin{bmatrix} 1 & 0 & 0 \\ 0 & 1 & 0 \\ 0 & 0 & 1 \end{bmatrix}.$$

It is obvious that there are no transitions from any state i to another state $j (i \neq j)$ except for each state itself, which implies three classes $\{0\}$, $\{1\}$ and $\{2\}$.

Example 5.3.3 Consider a Markov chain with the following transition probability matrix:

$$\mathbf{P} = \begin{matrix} 0 \\ 1 \\ 2 \end{matrix} \begin{bmatrix} 1 & 0 & 0 \\ 0 & \frac{1}{2} & \frac{1}{2} \\ \frac{1}{3} & \frac{1}{3} & \frac{1}{3} \end{bmatrix}.$$

It is obvious that $1 \leftrightarrow 2$ and $2 \rightarrow 0$. However, there is no transition $0 \rightarrow 2$. That is, there are two classes $\{0\}$ and $\{1, 2\}$.

We are now in a position to define an irreducible Markov chain.

Definition 5.3.1 For a Markov chain, if there exists only a single communication class, the Markov chain is said to be *irreducible*. That is, for an irreducible Markov chain, all the states communicate with each other.

We say that state i has *period* $d(i)$ if $d(i)$ is the greatest common divisor of $n \geq 1$ such that $p_{ii}^n > 0$. If $d(i) = 1$, then state i is *aperiodic*, and if $d(i) > 1$, then state i is *periodic*. We show the following theorem.

Theorem 5.3.2 If $i \leftrightarrow j$, then $d(i) = d(j)$.

Proof There exist integers m and n such that $p_{ij}^m > 0$ and $p_{ji}^n > 0$. If $p_{ii}^s > 0$, then

$$p_{jj}^{n+m} \geq p_{ji}^n p_{ij}^m > 0,$$

$$p_{jj}^{n+s+m} \geq p_{ji}^n p_{ii}^s p_{ij}^m > 0.$$

From the definition of the period, $d(j)$ divides both $n + m$ and $n + s + m$, and also the difference $n + s + m - (n + m) = s$, whenever $p_{ii}^s > 0$. That is, $d(j)$ divides $d(i)$. A similar argument by reversing i and j implies that $d(i)$ divides $d(j)$, which proves the theorem.

Example 5.3.4 Consider a Markov chain with the following transition probability matrix:

$$\mathbf{P} = \begin{matrix} 0 \\ 1 \\ 2 \end{matrix} \begin{bmatrix} 0 & 1 & 0 \\ 0 & 0 & 1 \\ 1 & 0 & 0 \end{bmatrix}.$$

It is obvious that $0 \to 1 \to 2 \to 0$, which implies an irreducible Markov chain. It is obvious that $p_{00}^3 = p_{00}^6 = \cdots = 1$, which implies that state 0 is periodic with $d(0) = 3$. Of course, $d(i) = 3$ $(i = 0,\ 1,\ 2)$ from Theorem 5.3.2.

Let us define the following probability:

$$f_{ij}^n = P\{X(n) = j,\ X(r) \neq j,\ r = 1,\ 2,\ \cdots,\ n-1 \mid X(0) = i\}, \quad (5.3.1)$$

for any $i,\ j$, which is the *first passage probability* from state i to state j with n steps, where we postulate $f_{ij}^0 = 0$ and $f_{ij}^1 = p_{ij}$. Noting that f_{ij}^n might be the probability mass function for fixed i and j, we define

$$f_{ij} = \sum_{n=1}^{\infty} f_{ij}^n \quad (5.3.2)$$

which is the *eventual transition probability* from state i to state j.

Definition 5.3.2 If $f_{ii} = 1$, then state i is said to be *recurrent*. If $f_{ii} < 1$, then state i is said to be *transient*.

We show the following necessary and sufficient conditions for identifying whether state i is recurrent or transient without proof.

Theorem 5.3.3 State i is recurrent if and only if

$$\sum_{n=1}^{\infty} p_{ii}^n = \infty. \quad (5.3.3)$$

State i is transient if and only if

$$\sum_{n=1}^{\infty} p_{ii}^n < \infty. \quad (5.3.4)$$

It is easy to understand that for any recurrent state i the process revisits state i infinitely, which implies Eq.(5.3.3).

Example 5.3.5 Consider a Markov chain with the following transition probability matrix:

$$\mathbf{P} = \begin{matrix} 0 \\ 1 \end{matrix} \begin{bmatrix} 1 & 0 \\ \frac{1}{2} & \frac{1}{2} \end{bmatrix}.$$

It is obvious that

$$\sum_{n=1}^{\infty} p_{00}^n = 1 + 1 + \cdots = \infty,$$

$$\sum_{n=1}^{\infty} p_{11}^n = \frac{1}{2} + (\frac{1}{2})^2 + \cdots = \frac{1/2}{1-1/2} = 1 < \infty,$$

which implies that state 0 is recurrent and state 1 is transient.

The next theorem shows that recurrence is a class property, just like periodicity.

Theorem 5.3.4 If state i is recurrent and $i \leftrightarrow j$, then state j is recurrent.

Proof There exist integers m and n such that $p_{ji}^m > 0$ and $p_{ij}^n > 0$. For any $s \geq 0$,

$$p_{jj}^{m+s+n} \geq p_{ji}^m p_{ii}^s p_{ij}^n.$$

Summing over s, we have

$$\sum_s p_{jj}^{m+s+n} \geq p_{ji}^m p_{ij}^n \sum_s p_{ii}^s = \infty,$$

which implies that state j is recurrent, which proves the theorem.

Corollary 5.3.1 An irreducible Markov chain is either recurrent or transient.

Example 5.3.6 (*One-Dimensional Random Walk*) Consider a Markov chain whose state space is the set of all integers and the probability of moving right is p, i.e.,

$$p_{i,i+1} = p, \; p_{i,i-1} = 1 - p = q \qquad (i = 0, \; \pm1, \; \pm2, \; \cdots),$$

where $0 < p < 1$. Noting that all states communicate, we have an irreducible Markov chain. Let us identify whether it is recurrent or transient. Let us calculate the transition probability that the process comes back to itself at $2n$ steps with n steps right and n steps left:

$$p_{00}^{2n} = \binom{2n}{n} p^n q^n = \frac{(2n)!}{(n!)^2}(pq)^n.$$

Applying *Stirling's formula*

$$n! \sim \sqrt{2\pi} n^{n+1/2} e^{-n},$$

we have

$$p_{00}^{2n} \sim \frac{(4pq)^n}{\sqrt{\pi n}}.$$

Summing over n implies

$$\sum_{n=1}^{\infty} \frac{(4pq)^n}{\sqrt{\pi n}},$$

which converges if $4pq = 4p(1-p) < 1$ and diverges if $4pq = 4p(1-p) = 1$, since equality holds for $4p(1-p) \leq 1$ if and only if $p = \frac{1}{2}$. Thus, the chain is recurrent if $p = \frac{1}{2}$ and transient if $p \neq \frac{1}{2}$.

When $p = \frac{1}{2}$, the chain is called a *one-dimensional symmetric random walk.* We can consider a two-dimensional symmetric random walk whose state space is the plane of all integers and the probability of moving up, down, right or left is 1/4 in each case. It is well-known that the process of the two-dimensional symmetric random walk is irreducible and recurrent (see Problem 5.10). However, all the higher-dimensional symmetric random walks greater than two are transient.

We say that state j is *absorbing* if j forms an equivalence class by itself. It is easy to verify that state j is absorbing if and only if $p_{jj} = 1$. It is sometimes called an *absorbing Markov chain* if there is at least one absorbing state.

The following theorem shows that a Markov chain forms several recurrent classes and a set of transient states.

Theorem 5.3.5 For a Markov chain, all states can be classified into several recurrent classes $C_1, C_2, \cdots$, and remaining states which are transient.

The proof is obvious and we omit it. Let $C_1, C_2, \cdots$ be all the recurrent classes and T be a set of all the remaining transient states for a Markov chain. Then all sets $C_1, C_2, \cdots, T$ are disjoint and exhaustive. Relabeling all the states suitably, we can rewrite as

$$\mathbf{P} = \begin{array}{c} C_1 \\ C_2 \\ \cdot \\ \cdot \\ \cdot \\ \\ T \end{array} \left[\begin{array}{ccccc:c} \mathbf{P}_1 & 0 & \cdot & \cdot & \cdot & 0 \\ 0 & \mathbf{P}_2 & \cdot & \cdot & \cdot & 0 \\ \cdot & \cdot & \cdot & & & \cdot \\ \cdot & \cdot & & \cdot & & \cdot \\ \cdot & \cdot & & & \cdot & \cdot \\ \hdashline \mathbf{R}_1 & \mathbf{R}_2 & \cdot & \cdot & \cdot & \mathbf{Q} \end{array} \right],$$

where the submatrices $\mathbf{P}_1, \mathbf{P}_2, \cdots$ are the transition probability matrices for respective recurrent classes $C_1, C_2, \cdots$, $\mathbf{Q}$ is a square matrix describing the transitions among all transient states for T, and $\mathbf{R}_1$, $\mathbf{R}_2$, $\cdots$ are (not necessarily square) matrices describing the transitions from all transient states to the corresponding recurrent classes C_1, C_2, $\cdots$.

Example 5.3.7 Consider a Markov chain with the following transition probability matrix:

$$\mathbf{P} = \begin{matrix} 0 \\ 1 \\ 2 \\ 3 \end{matrix} \begin{bmatrix} \frac{3}{4} & \frac{1}{4} & 0 & 0 \\ \frac{1}{3} & \frac{2}{3} & 0 & 0 \\ 0 & 0 & \frac{1}{2} & \frac{1}{2} \\ 0 & \frac{1}{3} & \frac{1}{3} & \frac{1}{3} \end{bmatrix}.$$

It is obvious that the chain has a recurrent class $\{0,\ 1\}$ and a set of transient states $\{2,\ 3\}$. Figure 5.3.1 shows a state transition diagram.

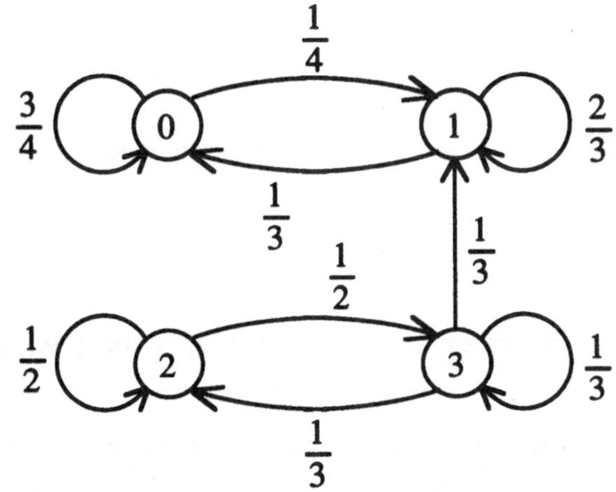

Fig. 5.3.1 A state transition diagram of Example 5.3.6.

5.4 Limiting Probabilities

Let us recall that f_{ij}^n is the probability mass function for fixed i and j. Let $N_j(t)$ denote the number of transitions into state j up to time t. For fixed j, if j is recurrent and $X(0) = j$, then the process $\{X(n),\ n = 0,\ 1,\ 2,\ \cdots\}$ is regarded as a discrete renewal process $\{N_j(t),\ t \geq 0\}$ with interarrival distribution $\{f_{jj}^n,\ n = 1,\ 2,\ \cdots\}$. For fixed i and j, if j is recurrent, $X(0) = i$, and $i \leftrightarrow j$, then the process is regarded as a discrete delayed renewal process with the first interarrival distribution $\{f_{ij}^n,\ n = 1,\ 2,\ \cdots\}$ and the successive interarrival distribution $\{f_{jj}^n,\ n = 1,\ 2,\ \cdots\}$.

We have classified all the states into recurrent or transient states by $f_{jj} = 1$ or $f_{jj} < 1$, respectively. It remains to classify all recurrent states.

Definition 5.4.1 For a Markov chain, all recurrent states are classified into *positive* (or *non-null*) *recurrent states* or *null recurrent states* by $\mu_j < \infty$ or $\mu_j = \infty$, respectively, where

$$\mu_j = \sum_{n=1}^{\infty} n f_{jj}^n \tag{5.4.1}$$

is the *mean recurrence time* for state j.

Recall that a Markov chain with state j being recurrent and $X(0) = j$ is regarded as a discrete renewal process with interarrival distribution $\{f_{jj}^n, \ n = 1, 2, \cdots\}$. Applying Theorem 4.3.5 (Blackwell's Theorem), we have the following limiting probability:

Theorem 5.4.1 If state j is recurrent and aperiodic, then

$$p_{jj}^n \longrightarrow \frac{1}{\mu_j} \tag{5.4.2}$$

as $n \to \infty$. And if state j is recurrent and periodic with period $d(j)$, then

$$p_{jj}^{nd(j)} \longrightarrow \frac{d(j)}{\mu_j} \tag{5.4.3}$$

as $n \to \infty$, where we interpret $1/\mu_j = 0$ if $\mu_j = \infty$ (i.e., if state j is null recurrent).

Proof If state j is recurrent and aperiodic, the interarrival distribution $\{f_{jj}^n, \ n = 1, 2, \cdots\}$ is not lattice. Applying Theorem 4.3.5 (ii), and assuming $\delta = 1$, we have

$$M(n) - M(n-1) = p_{jj}^n \longrightarrow \frac{1}{\mu_j} \tag{5.4.4}$$

as $n \to \infty$. If state j is recurrent and periodic with period $d(j)$, by applying Blackwell's Theorem 4.3.5 (ii), we have

$$M(nd(j)) - M((n-1)d(j)) = p_{jj}^{nd(j)} \longrightarrow \frac{d(j)}{\mu_j} \tag{5.4.5}$$

as $n \to \infty$, where $1/\mu_j = 0$ if state j is null recurrent (i.e., $\mu_j = \infty$).

The following corollary is obvious.

Corollary 5.4.1 If state j is transient, then

$$p_{jj}^n \longrightarrow 0 \tag{5.4.6}$$

as $n \to \infty$.

Let us consider the limiting behavior of p_{ij}^n as $n \to \infty$ in general. As shown above, we again focus on specified states i and j, where $X(0) = i$, a starting state. The Markov chain starts from state i, visits state j with probability mass function f_{ij}^n, and infinitely often revisits state j with probability mass function f_{jj}^n, which is regarded as a discrete delayed renewal process. Noting that f_{ij} denotes the eventual transition probability from state i to state j, we have the following theorem without proof:

Theorem 5.4.2 If state j is recurrent and aperiodic, then

$$p_{ij}^n \longrightarrow \frac{f_{ij}}{\mu_j} \tag{5.4.7}$$

as $n \to \infty$, where we interpret $1/\mu_j = 0$ if $\mu_j = \infty$.

We are now ready to derive the limiting probabilities for an irreducible Markov chain which is positive recurrent and aperiodic (such a Markov chain is sometimes called an *ergodic Markov chain*).

Theorem 5.4.3 If an irreducible Markov chain is positive recurrent and aperiodic, there exists the limiting probability

$$\lim_{n\to\infty} p_{ij}^n = \pi_j > 0 \qquad (i,\ j = 0,\ 1,\ 2,\ \cdots), \tag{5.4.8}$$

which is independent of the initial state i, where $\{\pi_j,\ j = 0,\ 1,\ 2,\ \cdots\}$ is a unique and positive solution to

$$\pi_j = \sum_{i=0}^{\infty} \pi_i p_{ij} \qquad (j = 0,\ 1,\ 2,\ \cdots), \tag{5.4.9}$$

$$\sum_{j=0}^{\infty} \pi_j = 1, \tag{5.4.10}$$

and is called a *stationary distribution* for a Markov chain.

The proof needs much elaboration and we omit it. However, we interpret the theorem above. First, if an irreducible Markov chain is positive recurrent and aperiodic, we can prove Eq.(5.4.8) from Theorem 5.4.2 since $f_{ij} = 1$ for all i and j. Secondly, we briefly show how to derive Eqs.(5.4.9) and (5.4.10). As shown in Eqs.(5.2.12) and (5.2.14), we have

$$\pi_j(n) = \sum_{i=0}^{\infty} \pi_i(n-1)p_{ij} \qquad (j = 0,\ 1,\ 2,\ \cdots), \tag{5.4.11}$$

and

$$\sum_{j=0}^{\infty} \pi_j(n) = 1. \tag{5.4.12}$$

Noting that

$$\lim_{n\to\infty} p_{ij}^n = \lim_{n\to\infty} \pi_j(n) = \pi_j > 0 \qquad (j = 0,\ 1,\ 2,\ \cdots), \tag{5.4.13}$$

and letting $n \to \infty$ in Eqs.(5.4.11) and (5.4.12), we have Eqs.(5.4.9) and (5.4.10).

Example 5.4.1 (*Example 5.2.1*) The stationary distribution $\pi = [\pi_0\ \pi_1]$ is given by

$$\begin{cases} \pi_0 = 0.8\pi_0 + 0.3\pi_1 \\ \pi_1 = 0.2\pi_0 + 0.7\pi_1 \\ \pi_0 + \pi_1 = 1. \end{cases}$$

Note that the first two equations are the same since $2\pi_0 = 3\pi_1$ by simplification. Solving a set of simultaneous equations $2\pi_0 = 3\pi_1$ and $\pi_0 + \pi_1 = 1$ implies

$$\pi_0 = 0.6, \ \pi_1 = 0.4.$$

Example 5.4.2 (*Example 5.2.2*) The stationary distribution $\pi = [\pi_0\ \pi_1]$ is given by

$$\begin{cases} \pi_0 = 0.4\pi_0 + 0.9\pi_1 \\ \pi_1 = 0.6\pi_0 + 0.1\pi_1 \\ \pi_0 + \pi_1 = 1, \end{cases}$$

whose solution is given by

$$\pi_0 = 0.6, \ \pi_1 = 0.4.$$

Figures 5.2.2 and 5.2.3 show that the transition probabilities convergé toward stationary distribution as $n \to \infty$. It is surprising that both examples have the same stationary distribution.

Example 5.4.3 (*Example 5.3.7*) The chain is not irreducible, and has a recurrent class {0, 1} and a set of transient states {2, 3}. A submatrix of the recurrent class C_1 is given by

$$\mathbf{P}_1 = \begin{bmatrix} \frac{3}{4} & \frac{1}{4} \\ \frac{1}{3} & \frac{2}{3} \end{bmatrix},$$

and the stationary distribution is given by

$$\begin{cases} \pi_0 = \frac{3}{4}\pi_0 + \frac{1}{3}\pi_1 \\ \pi_1 = \frac{1}{4}\pi_0 + \frac{2}{3}\pi_1 \\ \pi_0 + \pi_1 = 1. \end{cases}$$

As in Example 5.4.1, we have a unique and positive solution:

$$\pi_0 = \frac{4}{7}, \ \pi_1 = \frac{3}{7}.$$

If the chain starts from $i = 0, \ 1 \in C_1$, we have the limiting probabilities:

$$\lim_{n\to\infty} p_{i0}^n = \pi_0 = \frac{4}{7} \qquad (i = 0, \ 1),$$

$$\lim_{n\to\infty} p_{i1}^n = \pi_1 = \frac{3}{7} \qquad (i = 0, \ 1).$$

However, if the chain starts from $i = 2, \ 3 \in T$, we apply Theorem 5.4.2 and have

$$\lim_{n\to\infty} p_{i0}^n = \frac{f_{i0}}{\mu_0} = \frac{4}{7} f_{i0} \qquad (i = 2, \ 3),$$

$$\lim_{n\to\infty} p_{i1}^n = \frac{f_{i1}}{\mu_1} = \frac{3}{7} f_{i1} \qquad (i = 2, \ 3).$$

We will show how to derive f_{ij} $(i \in T, \ j \in C_1)$ in the following section. Of course, we have

$$\lim_{n\to\infty} p_{ij}^n = 0 \qquad (i, \ j \in T)$$

from Corollary 5.4.1.

Example 5.4.4 (*Random Walk with Reflecting Barrier*) A random walk with reflective barrier is a Markov chain whose state space is the set of non-negative integers and whose transition probability is given by

$$p_{i,i+1} = p_i, \ p_{i,i-1} = 1 - p_i = q_i \qquad (i = 1, \ 2, \ 3, \ \cdots)$$

where $p_0 = 1$ (i.e., $p_{01} = 1$). The transition probability matrix is given by

$$\mathbf{P} = \begin{bmatrix} 0 & 1 & 0 & 0 & \cdots \\ q_1 & 0 & p_1 & 0 & \cdots \\ 0 & q_2 & 0 & p_2 & \cdots \\ 0 & 0 & q_3 & 0 & \cdots \\ \vdots & \vdots & \vdots & \vdots & \ddots \end{bmatrix}.$$

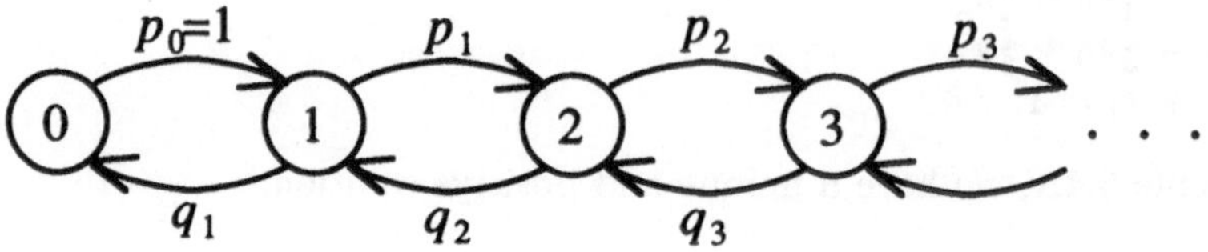

Fig. 5.4.1 A state transition diagram of the random walk.

Figure 5.4.1 shows a state transition diagram of the random walk. We observe that the chain is irreducible, recurrent, and periodic with period 2. Let us derive the stationary distribution. Assuming that $p_{-1} = 0$, $p_0 = 1$, and $\pi_{-1} = 0$, Eq.(5.4.9) is given by

$$\pi_j = \sum_{i=0}^{\infty} \pi_i p_{ij} = \pi_{j-1}p_{j-1} + \pi_{j+1}q_{j+1} \qquad (j = 0,\ 1,\ 2,\ \cdots).$$

Noting that $p_0 = 1$ and $p_j + q_j = 1$, we can rewrite

$$\begin{aligned} \pi_1 q_1 &= \pi_0 p_0, \\ \pi_2 q_2 - \pi_1 q_1 &= \pi_1 p_1 - \pi_0 p_0, \\ \pi_3 q_3 - \pi_2 q_2 &= \pi_2 p_2 - \pi_1 p_1, \\ &\vdots \\ \pi_{j+1} q_{j+1} - \pi_j q_j &= \pi_j p_j - \pi_{j-1} p_{j-1}. \end{aligned}$$

Summing over all left- and right-hand sides, we have

$$\pi_{j+1} q_{j+1} = \pi_j p_j \qquad (j = 0,\ 1,\ \cdots)$$

which yields

$$\pi_{j+1} = \pi_0 \prod_{k=0}^{j} \frac{p_k}{q_{k+1}} \qquad (j = 0,\ 1,\ 2,\ \cdots).$$

The necessary and sufficient condition that the chain is positive recurrent is that there exists $\pi_0 > 0$ such that

$$\sum_{j=0}^{\infty} \pi_j = \pi_0 \left[1 + \sum_{j=0}^{\infty} \prod_{k=0}^{j} \frac{p_k}{q_{k+1}} \right] = 1.$$

That is, the condition is given by

$$\sum_{j=0}^{\infty} \prod_{k=0}^{j} \frac{p_k}{q_{k+1}} < \infty.$$

In particular, if $p_i = p$ $(q_i = q)$ $(i = 1,\ 2,\ 3,\ \cdots)$, we have the necessary and sufficient condition that $p < q$ (i.e., $p < 1/2$). Then the stationary distribution is given by

$$\pi_0 = \frac{q-p}{2q},\ \pi_j = \frac{q-p}{2q} \cdot \frac{1}{q} \left(\frac{p}{q} \right)^{j-1} \qquad (j = 1,\ 2,\ \cdots).$$

See the limiting probabilities in Problem 5.16.

5.5 Finite-State Markov Chains

In this section we restrict ourselves to a Markov chain with finite states, which is called a *finite-state Markov chain.* Let the state space be $i = 0,\ 1,\ 2,\ \cdots, N$, unless otherwise specified. That is, we consider a finite-state Markov chain with $N + 1$ states. First, we give the following theorem and corollary:

Theorem 5.5.1 For a finite-state Markov chain:

(i) There are no null recurrent states.

(ii) Not all states are transient.

Proof To prove **(ii)**, if all states are transient, $\lim_{n \to \infty} p_{ij}^n = 0$ for all j, which contradicts $\sum_{j=0}^{N} p_{ij}^n = 1$ for all n. To prove **(i)**, if a state j is null recurrent and the state space is finite, then the mean recurrence time μ_j must be finite, which contradicts null recurrence. That is, the recurrent state is positive recurrent.

Corollary 5.5.1 An irreducible finite-state Markov chain is positive recurrent.

The proof is obvious. If the irreducible Markov chain is aperiodic, there exists a unique and positive stationary distribution which is the limiting distribution.

We develop the limiting probabilities for a finite-state Markov chain in general. Applying Theorem 5.3.5 for a finite-state Markov chain, we can relabel all the states and rewrite the transition probability matrix in the following from:

$$\mathbf{P} = \begin{array}{c} C_1 \\ C_2 \\ \cdot \\ \cdot \\ \cdot \\ C_m \\ T \end{array} \left[\begin{array}{cccccc:c} \mathbf{P}_1 & 0 & \cdot & \cdot & \cdot & 0 & 0 \\ 0 & \mathbf{P}_2 & \cdot & \cdot & \cdot & 0 & 0 \\ \cdot & \cdot & & & & \cdot & \cdot \\ \cdot & \cdot & & \cdot & & \cdot & \cdot \\ \cdot & \cdot & & & & \cdot & \cdot \\ 0 & 0 & \cdot & \cdot & \cdot & \mathbf{P}_m & 0 \\ \hdashline \mathbf{R}_1 & \mathbf{R}_2 & \cdot & \cdot & \cdot & \mathbf{R}_m & \mathbf{Q} \end{array}\right], \tag{5.5.1}$$

where $C_1,\ C_2,\ \cdots C_m$ are all the sets of positive recurrent classes and T is a set of the remaining transient states. As shown in Example 5.4.3, we can obtain the limiting probabilities $p_{ij}^{\infty} = \lim_{n\to\infty} p_{ij}^n$ for all states. Assuming that all positive recurrent states are aperiodic, we summarize the following:

$$p_{ij}^{\infty} = 1/\mu_j \qquad (i,\ j \in C_k\ ; k = 1,\ 2,\ \cdots, m), \tag{5.5.2}$$

$$p_{ij}^{\infty} = f_{ij}/\mu_j \qquad (i \in T,\ j \in C_k\ ; k = 1,\ 2,\ \cdots, m), \tag{5.5.3}$$

$$p_{ij}^{\infty} = 0 \qquad (i \in C_k,\ j \in C_l\ ; k \neq l), \tag{5.5.4}$$

$$p_{ij}^{\infty} = 0 \qquad (i,\ j \in T),$$

$$p_{ij}^{\infty} = 0 \qquad (i \in C_k\ ; k = 1,\ 2,\ \cdots, m\ ; j \in T). \tag{5.5.5}$$

In particular, we apply Theorem 5.4.2 for deriving Eq.(5.5.3). The above results are not perfect since we have to give the expressions f_{ij} for $i \in T$ and $j \in C_k$ ($k = 1,\ 2,\ \cdots, m$).

Example 5.5.1 Consider a finite-state Markov chain with the following transition probability matrix:

$$\mathbf{P} = \begin{array}{c} 0 \\ 1 \\ 2 \\ 3 \\ 4 \end{array} \left[\begin{array}{cc:c:cc} 1/4 & 3/4 & 0 & 0 & 0 \\ 1/2 & 1/2 & 0 & 0 & 0 \\ \hdashline 0 & 0 & 1 & 0 & 0 \\ \hdashline 1/2 & 0 & 0 & 1/4 & 1/4 \\ 0 & 1/3 & 1/3 & 0 & 1/3 \end{array}\right].$$

A state transition diagram is shown in Fig. 5.5.1. It is obvious that the chain has two positive recurrent classes $C_1 = \{0,\ 1\}$ and $C_2 = \{2\}$, and a set of remaining transient states $T = \{3, 4\}$.

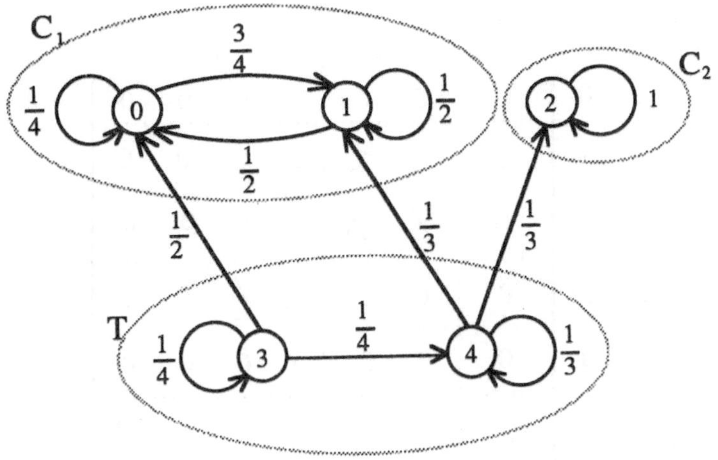

Fig. 5.5.1 A state transition diagram of Example 5.5.1.

Example 5.5.2 Consider a finite-state Markov chain with the following transition probability matrix:

$$\mathbf{P} = \begin{matrix} 0' \\ 1' \\ 2' \\ 3' \\ 4' \\ 5' \\ 6' \end{matrix} \begin{bmatrix} 0 & 0 & 0 & 0 & 1 & 0 & 0 \\ 0 & 0 & 1/3 & 1/3 & 0 & 0 & 1/3 \\ 0 & 0 & 1/2 & 0 & 0 & 1/2 & 0 \\ 0 & 0 & 0 & 1 & 0 & 0 & 0 \\ 1/2 & 0 & 0 & 0 & 1/2 & 0 & 0 \\ 0 & 0 & 3/4 & 0 & 0 & 1/4 & 0 \\ 0 & 1/2 & 0 & 0 & 1/2 & 0 & 0 \end{bmatrix}.$$

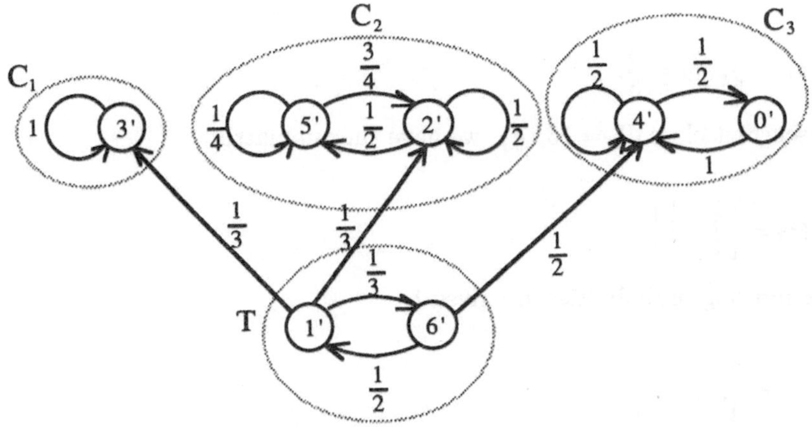

Fig. 5.5.2 A state transition diagram of Example 5.5.2.

The relabeling of new states such that $3' \Rightarrow 0$, $5' \Rightarrow 1$, $2' \Rightarrow 2$, $4' \Rightarrow 3$, $0' \Rightarrow 4$, $6' \Rightarrow 5$, $1' \Rightarrow 6$, allows us to produce the following transition probability matrix:

$$\mathbf{P} = \begin{array}{c} 0 \\ 1 \\ 2 \\ 3 \\ 4 \\ 5 \\ 6 \end{array} \left[\begin{array}{c|cc|cc|cc} 1 & 0 & 0 & 0 & 0 & 0 & 0 \\ \hline 0 & 1/4 & 3/4 & 0 & 0 & 0 & 0 \\ 0 & 1/2 & 1/2 & 0 & 0 & 0 & 0 \\ \hline 0 & 0 & 0 & 1/2 & 1/2 & 0 & 0 \\ 0 & 0 & 0 & 1 & 0 & 0 & 0 \\ \hline 0 & 0 & 0 & 1/2 & 0 & 0 & 1/2 \\ 1/3 & 0 & 1/3 & 0 & 0 & 1/3 & 0 \end{array} \right].$$

That is, we have a chain that has three positive recurrent classes $C_1 = \{0\}$, $C_2 = \{1,\ 2\}$, $C_3 = \{3,\ 4\}$, and a set of remaining transient states $T = \{5,\ 6\}$ for relabeled states. The limiting probabilities for a recurrent class can be easily obtained by Theorem 5.4.3. For a recurrent class $C_2 = \{1,\ 2\}$, we have the submatrix:

$$\mathbf{P}_2 = \begin{bmatrix} 1/4 & 3/4 \\ 1/2 & 1/2 \end{bmatrix}$$

and the limiting probabilities are given by

$$\pi_1 = \frac{2}{5},\ \pi_2 = \frac{3}{5}.$$

For a recurrent class $C_3 = \{3,\ 4\}$, we have the submatrix

$$\mathbf{P}_3 = \begin{bmatrix} \frac{1}{2} & \frac{1}{2} \\ 1 & 0 \end{bmatrix}$$

and the limiting probabilities are given by

$$\pi_3 = \frac{2}{3},\ \pi_4 = \frac{1}{3}.$$

We have to obtain f_{ij} for $i \in T$ and $j \in C_k$ $(k = 1,\ 2,\ 3)$.

We are now in a position to show the following theorem:

Theorem 5.5.2 For a finite-state Markov chain, the eventual transition probability f_{ij} from state $i \in T$ to state $j \in C_k$ $(k = 1, 2, \cdots, m)$ satisfies

$$f_{ij} = \sum_{l \in C_k} p_{il} + \sum_{l \in T} p_{il} f_{lj} \qquad (i \in T,\ j \in C_k), \tag{5.5.6}$$

and in a matrix form

$$[f_{ij}] = [\mathbf{I} - \mathbf{Q}]^{-1} \mathbf{R}_k \cdot \mathbf{1} \tag{5.5.7}$$

where $\mathbf{1}$ is a column vector of all the components' unity.

Proof The eventual transition probability f_{ij} from state $i \in T$ to state $j \in C_k$ has two possibilities: One is to move to a state $l \in C_k$ directly and another is to move to a state $l \in T$ including itself. The latter obeys the eventual transition probability f_{lj}. Thus we have Eq.(5.5.6). In a matrix of Eq.(5.5.6), we have

$$[f_{ij}] = \mathbf{R}_k \cdot \mathbf{1} + \mathbf{Q} \cdot [f_{ij}]$$

whose solution is Eq.(5.5.7). Note that the inverse $[\mathbf{I} - \mathbf{Q}]^{-1}$ exists since $\mathbf{Q}$ represents the transition probability matrix among all transient states and no states remain forever in set T.

Example 5.5.3 (*Example 5.5.2*) Let us derive f_{ij} for $i \in T = \{5, 6\}$ and $j \in C_k$ $(k = 1, 2, 3)$. First,

$$[\mathbf{I} - \mathbf{Q}] = \begin{bmatrix} 1 & -\frac{1}{2} \\ -\frac{1}{3} & 1 \end{bmatrix}$$

and the inverse is given by

$$[\mathbf{I} - \mathbf{Q}]^{-1} = \begin{bmatrix} \frac{6}{5} & \frac{3}{5} \\ \frac{2}{5} & \frac{6}{5} \end{bmatrix}.$$

Applying Theorem 5.5.2 implies

$$[\mathbf{I} - \mathbf{Q}]^{-1} \mathbf{R}_1 \cdot \mathbf{1} = \begin{bmatrix} \frac{6}{5} & \frac{3}{5} \\ \frac{2}{5} & \frac{6}{5} \end{bmatrix} \begin{bmatrix} 0 \\ \frac{1}{3} \end{bmatrix} [1] = \begin{bmatrix} \frac{1}{5} \\ \frac{2}{5} \end{bmatrix},$$

$$[\mathbf{I} - \mathbf{Q}]^{-1} \mathbf{R}_2 \cdot \mathbf{1} = \begin{bmatrix} \frac{6}{5} & \frac{3}{5} \\ \frac{2}{5} & \frac{6}{5} \end{bmatrix} \begin{bmatrix} 0 & 0 \\ 0 & \frac{1}{3} \end{bmatrix} \begin{bmatrix} 1 \\ 1 \end{bmatrix} = \begin{bmatrix} \frac{1}{5} \\ \frac{2}{5} \end{bmatrix},$$

$$[\mathbf{I}-\mathbf{Q}]^{-1}\mathbf{R}_3\cdot\mathbf{1}=\begin{bmatrix}\frac{6}{5} & \frac{3}{5}\\ \frac{2}{5} & \frac{6}{5}\end{bmatrix}\begin{bmatrix}\frac{1}{2} & 0\\ 0 & 0\end{bmatrix}\begin{bmatrix}1\\ 1\end{bmatrix}=\begin{bmatrix}\frac{3}{5}\\ \frac{1}{5}\end{bmatrix}.$$

That is,

$$f_{50}=\frac{1}{5},\ f_{60}=\frac{2}{5},$$

$$f_{5j}=\frac{1}{5},\ f_{6j}=\frac{2}{5}\qquad (j=1,\ 2),$$

$$f_{5j}=\frac{3}{5},\ f_{6j}=\frac{1}{5}\qquad (j=3,\ 4).$$

Applying Eqs.(5.5.2) - (5.5.5) yields

$$\mathbf{P}^{\infty}=\lim_{n\to\infty}\mathbf{P}^{n}=\begin{array}{c} 0\\ 1\\ 2\\ 3\\ 4\\ 5\\ 6\end{array}\left[\begin{array}{c:cc:cc:cc} 1 & 0 & 0 & 0 & 0 & 0 & 0\\ \hdashline 0 & 2/5 & 3/5 & 0 & 0 & 0 & 0\\ 0 & 2/5 & 3/5 & 0 & 0 & 0 & 0\\ \hdashline 0 & 0 & 0 & 2/3 & 1/3 & 0 & 0\\ 0 & 0 & 0 & 2/3 & 1/3 & 0 & 0\\ \hdashline 1/5 & 2/25 & 3/25 & 2/5 & 1/5 & 0 & 0\\ 2/5 & 4/25 & 6/25 & 2/15 & 1/15 & 0 & 0\end{array}\right],$$

where we can verify that

$$\sum_{j=0}^{4}p_{ij}^{\infty}=1\qquad (i=0,\ 1,\ 2,\ 3,\ 4,\ 5,\ 6).$$

5.6 Problems 5

5.1 Consider a Markov chain with the following transition probability matrix:

$$\mathbf{P} = \begin{matrix} 0 \\ 1 \end{matrix} \begin{bmatrix} 0.6 & 0.4 \\ 0.8 & 0.2 \end{bmatrix}.$$

Calculate the n-step distribution $\pi(n) = [\pi_0(n)\ \pi_1(n)]$ $(n = 2, 4, 8)$ by assuming the following initial distribution:

(i) $\pi(0) = [1\ 0]$.

(ii) $\pi(0) = [0.5\ 0.5]$.

(iii) $\pi(0) = [2/3\ 1/3]$.

5.2 Consider an absorbing Markov chain with the following transition probability matrix:

$$\mathbf{P} = \begin{matrix} 0 \\ 1 \\ 2 \end{matrix} \begin{bmatrix} 1 & 0 & 0 \\ 1/3 & 2/3 & 0 \\ 0 & 1/2 & 1/2 \end{bmatrix}.$$

(i) Show a state transition diagram.

(ii) Calculate $\mathbf{P}^2$, $\mathbf{P}^3$, and $\mathbf{P}^4$.

(iii) Calculate $\mathbf{P}^n$ $(n = 1, 2, \cdots)$, in general, by considering a state transition diagram.

5.3 For a finite-state Markov chain with $(N+1)$ states, verify that, if $i \to j$ and $i \neq j$, state j is accessible from state i within N steps.

5.4 Consider two urns A and B, where, initially, A has two *white* balls and B has two *black* balls. An experiment is performed in which a ball is selected at random (with equal likelihood) from each urn and placed in the other urn. Repeat this experiment. Assuming that state i $(i = 0, 1, 2)$ denotes that A has i black balls, show the following:

(i) Verify that process $\{X(n), n = 0, 1, 2, \cdots\}$ is a discrete-time Markov chain.

(ii) Derive a transition probability matrix $\mathbf{P}$.

(iii) Classify all the states of the chain (i.e., identify that each state is recurrent or transient, periodic or aperiodic, and so on).

5.5 (*Continuation*) Consider two urns A and B, where, initially, A has N *white* balls and B has N *black* balls. The same experiment is performed as in Problem 5.4. Verify that

(i) The process $\{X(n),\ n = 0, 1, 2, \cdots\}$ is a Markov chain, where $X(n) = i$ $(i = 0, 1, 2, \cdots, N)$ denotes that A has i black balls.

(ii) Derive a transition probability matrix.

(iii) Classify all the states of the chain (i.e., identify that each state is recurrent or transient, periodic or aperiodic, and so on).

5.6 Consider a two-state Markov chain with the following transition probability matrix $\mathbf{P}$:

$$\mathbf{P} = \begin{matrix} 0 \\ 1 \end{matrix} \begin{bmatrix} 1-a & a \\ b & 1-b \end{bmatrix},$$

where $0 < a < 1,\ 0 < b < 1$ and $| 1 - a - b | < 1$.

(i) Derive two distinct eigenvalues $\lambda_0,\ \lambda_1$ of $\mathbf{P}$, i.e., solve $| \mathbf{P} - \lambda \mathbf{I} | = 0$, where $\mathbf{I}$ is an identity matrix.

(ii) Let $\mathbf{x}_i$ and $\mathbf{y}_i'$ be the left and right eigenvectors of $\mathbf{P}$ corresponding to the eigenvalues λ_i $(i = 0, 1)$, where $\mathbf{x}_i$ and $\mathbf{y}_i$ are column vectors and $\mathbf{y}_i'$ is a row vector. That is, $\mathbf{P}\mathbf{x}_i = \lambda_i \mathbf{x}_i,\ \mathbf{y}_i'\mathbf{P} = \lambda_i \mathbf{y}_i'$ $(i = 0, 1)$. Derive the left and right eigenvectors of $\mathbf{P}$ $(i = 0, 1)$, and choose a suitable multiplication factor such that $\mathbf{y}_i'\mathbf{x}_i = 1$ $(i = 0, 1)$.

(iii) Derive $\mathbf{H} = [\mathbf{x}_0,\ \mathbf{x}_1]$ and its inverse $\mathbf{H}^{-1} = \begin{bmatrix} \mathbf{y}_0' \\ \mathbf{y}_1' \end{bmatrix}$.

(iv) Verify that $\mathbf{P} = \mathbf{H}\mathbf{\Lambda}\mathbf{H}^{-1}$, where $\mathbf{\Lambda} = \begin{bmatrix} \lambda_0 & 0 \\ 0 & \lambda_1 \end{bmatrix}$.

(v) Derive the n-step transition probability matrix:

$$\mathbf{P}^n = \frac{1}{a+b} \begin{bmatrix} b & a \\ b & a \end{bmatrix} + \frac{(1-a-b)^n}{a+b} \begin{bmatrix} a & -a \\ -b & b \end{bmatrix}.$$

5.7 (*Continuation*) Verify that the n-step transition probability matrix is given by

$$\mathbf{P}^n = \frac{1}{a+b} \begin{bmatrix} b & a \\ b & a \end{bmatrix} + \frac{(1-a-b)^n}{a+b} \begin{bmatrix} a & -a \\ -b & b \end{bmatrix},$$

by the techniques of mathematical induction.

5.8 Consider the Markov chains with the following transition probability matrices:

$$\mathbf{P}_1 = \begin{bmatrix} 0 & 1 & 0 \\ 0 & 0 & 1 \\ 0.5 & 0.5 & 0 \end{bmatrix}, \quad \mathbf{P}_2 = \begin{bmatrix} 0.2 & 0.5 & 0.3 \\ 0 & 1 & 0 \\ 0.9 & 0.1 & 0 \end{bmatrix},$$

$$\mathbf{P}_3 = \begin{bmatrix} 0.4 & 0.6 & 0 & 0 \\ 0.2 & 0.8 & 0 & 0 \\ 0 & 0 & 1 & 0 \\ 0 & 0 & 0.5 & 0.5 \end{bmatrix}.$$

(i) Show a state transition diagram for each Markov chain.

(ii) Classify all the states, i.e., identify all the equivalence classes and a set of transient states, and whether each state is recurrent or transient.

5.9 Consider the Markov chains with the following transition probability matrices:

$$\mathbf{P}_1 = \begin{bmatrix} 0 & 1 & 0 \\ 0 & 0 & 1 \\ 1 & 0 & 0 \end{bmatrix}, \quad \mathbf{P}_2 = \begin{bmatrix} 0 & 1 & 0 \\ 0 & 0 & 1 \\ 0.5 & 0.5 & 0 \end{bmatrix},$$

$$\mathbf{P}_3 = \begin{bmatrix} 0 & 0 & 0 & 0.4 & 0.6 \\ 0 & 0 & 0 & 0.2 & 0.8 \\ 0 & 0 & 0 & 0 & 1 \\ 0.5 & 0.5 & 0 & 0 & 0 \\ 0 & 1 & 0 & 0 & 0 \end{bmatrix}.$$

Show that each Markov chain is irreducible, and derive its period.

5.10 (*A two-dimensional symmetric random walk*)

Consider a two-dimensional symmetric random walk on all the integers (i, j) $(i, j = 0, \pm 1, \pm 2, \cdots)$ in the infinite plane. Let the probabilities of moving to the right, left, up and down all be equal to $1/4$.

(i) Show that the Markov chain of this random walk is irreducible and periodic with period 2.

(ii) Verify that

$$\begin{aligned} p_{00}^{2n} &= \sum_{\substack{i,j \\ i+j=n}} \frac{(2n)!}{i!\,i!\,j!\,j!} \left(\frac{1}{4}\right)^{2n} \\ &= \left(\frac{1}{4}\right)^{2n} \binom{2n}{n} \sum_{r=0}^{n} \binom{n}{r} \binom{n}{n-r} \\ &= \left(\frac{1}{4}\right)^{2n} \binom{2n}{n}^2 \end{aligned}$$

(Hint: refer to Problem 1.8).

(iii) Verify that, by using Stirling's formula,

$$p_{00}^{2n} \sim \frac{1}{\pi n}$$

and

$$\sum_{n=0}^{\infty} p_{00}^{2n} = \infty$$

i.e., the Markov chain is recurrent.

5.11 Consider the Markov chains with the following transition probability matrices:

$$\mathbf{P}_1 = \begin{bmatrix} 2/3 & 1/3 \\ 1/3 & 2/3 \end{bmatrix}, \ \mathbf{P}_2 = \begin{bmatrix} 1/2 & 1/2 \\ 1/2 & 1/2 \end{bmatrix}, \ \mathbf{P}_3 = \begin{bmatrix} 0 & 1 \\ 1 & 0 \end{bmatrix}.$$

(i) Show that each Markov chain is irreducible.

(ii) Derive the period of each Markov chain.

(iii) Derive the stationary distribution for each Markov chain, i.e., derive the right eigenvector $\pi = \pi \mathbf{P}_i$ $(i = 1, 2, 3)$ such that $\pi_0 + \pi_1 = 1$.

(iv) Calculate the limiting transition probability matrices for Markov chains with $\mathbf{P}_1$ and $\mathbf{P}_2$.

(v) Calculate the following limiting probability matrix for a Markov chain with $\mathbf{P}_3$:

$$\lim_{n\to\infty} \frac{1}{n} \sum_{i=1}^{n} \mathbf{P}_3^i.$$

5.12 Calculate the limiting transition probability matrix for a Markov chain with the following transition probability matrix:

$$\mathbf{P} = \begin{bmatrix} 0 & 1 & 0 & 0 & 0 \\ 0.5 & 0.5 & 0 & 0 & 0 \\ 0 & 0 & 1 & 0 & 0 \\ 0 & 0 & 0 & 0.4 & 0.6 \\ 0 & 0 & 0 & 0.9 & 0.1 \end{bmatrix}.$$

5.13 Consider a finite-state Markov chain with the following transition probability matrix:

$$\mathbf{P} = \begin{matrix} 0' \\ 1' \\ 2' \\ 3' \\ 4' \\ 5' \end{matrix} \begin{bmatrix} 1/4 & 1/2 & 0 & 0 & 1/4 & 0 \\ 1/4 & 0 & 0 & 3/4 & 0 & 0 \\ 0 & 0 & 1/3 & 0 & 2/3 & 0 \\ 0 & 0 & 0 & 2/3 & 0 & 1/3 \\ 0 & 0 & 1/2 & 0 & 1/2 & 0 \\ 0 & 0 & 0 & 1/2 & 0 & 1/2 \end{bmatrix}.$$

Relabel all states and rewrite the transition probability matrix as in Eq. (5.5.1). Derive the limiting transition probability matrix $\mathbf{P}^\infty = \lim_{n\to\infty} \mathbf{P}^n$ for the Markov chain with relabeled states. (Hint: refer to Examples 5.5.2 and 5.5.3).

5.14 For a finite-state Markov chain with $(N+1)$ states, if we assume that

$$\sum_{i=0}^{N} p_{ij} = 1 \qquad (j = 0, 1, 2, \cdots, N),$$

then such a transition probability matrix is called a *doubly stochastic matrix* since each row sum and column sum is a unity (i.e., doubly stochastic). Verify that the limiting probability is given by

$$\pi_j = \lim_{n\to\infty} p_{ij}^n = \frac{1}{N+1} \quad (i, j = 0, 1, \cdots, N),$$

if the Markov chain with a doubly stochastic matrix is irreducible, recurrent, and aperiodic.

5.15 A particle can move clockwise with probability p $(0 < p < 1)$ and counterclockwise with $q = 1 - p$ on each state i $(i = 0, 1, 2, 3)$ on a circle shown below. That is, we consider a random walk on a circle. Derive the transition probability matrix for such a Markov chain, and identify all its states.

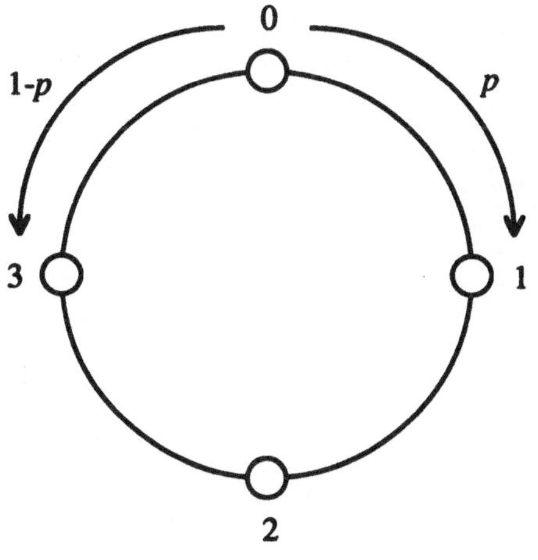

5.16 (*Example 5.4.4*) Consider a random walk with reflecting barrier.

(i) Verify that the Markov chain is irreducible and periodic with period 2.

(ii) Derive the limiting transition probability matrices:

$$\lim_{n\to\infty} \mathbf{P}^{2n}, \quad \lim_{n\to\infty} \mathbf{P}^{2n+1}$$

by using the stationary distribution in Example 5.4.4.

Chapter 6

Continuous-Time Markov Chains

6.1 Introduction

In this chapter we again develop continuous-time stochastic processes having Markov properties. The continuous-time Markov chain is defined in the following:

Definition 6.1.1 Let $\{X(t),\ t \geq 0\}$ be a continuous-time stochastic process with state space $i = 0, 1, 2, \cdots$, unless otherwise specified. If

$$P\{X(t) = x \mid X(t_1) = x_1, X(t_2) = x_2, \cdots, X(t_n) = x_n\}$$

$$= P\{X(t) = x \mid X(t_n) = x_n\}, \tag{6.1.1}$$

for all $0 \leq t_1 < t_2 < \cdots < t_n < t$, then the process is called a *continuous-time Markov chain.* For any $t \geq 0, s \geq 0$,

$$P_{ij}(t) = P\{X(t+s) = j \mid X(s) = i\} \tag{6.1.2}$$

is called a *transition probability,* where we assume that $P_{ij}(t)$ in Eq.(6.1.2) is independent of time s, i.e., the process is *stationary.*

Throughout this chapter we develop the continuous-time Markov chains with stationary transition probabilities. Naturally, our interest is to derive the explicit expressions for $P_{ij}(t)$ in Eq.(6.1.2) by assuming a specified model. Similar to the discussions in Section 5.2, we derive the Chapman-Kolmogorov equation for a continuous-time Markov chain in general. The transition probability $P_{ij}(t+s)$

can be calculated by summing over the entire intermediate state k at time t and moving to state j from k at the remaining time s (see Fig. 6.1.1). That is,

$$P_{ij}(t+s) = \sum_{k=0}^{\infty} P_{ik}(t)P_{kj}(s) \tag{6.1.3}$$

for any i and j. Equation (6.1.3) is called the *Chapman-Kolmogorov equation* for the continuous-time Markov chain. However, we cannot proceed to derive the explicit solution for $P_{ij}(t)$ in general. We have to assume some conditions to develop the process in detail.

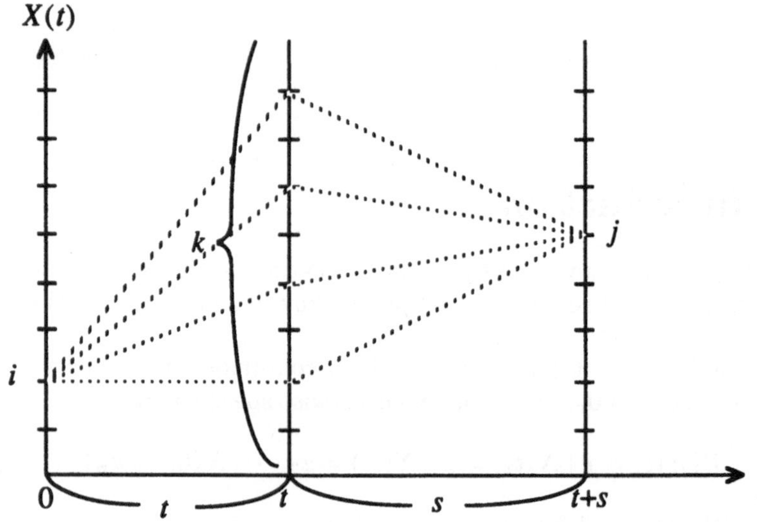

Fig. 6.1.1 An interpretation of Chapman-Kolmogorov's equation.

The continuous-time Markov chain plays an important role in both theory and application in many fields. These include queueing and inventory theories, population growth, biology, economics, engineering systems, and the social sciences. For example, the main knowledge of queueing theory is a field of application for continuous-time Markov chains.

In this chapter we develop continuous-time Markov chains. Sections 6.2 and 6.3 introduce the pure birth processes and pure death processes, respectively. Section 6.4 discusses the birth and death processes. Section 6.5 introduces the finite-state continuous-time Markov chains.

6.2 Pure Birth Processes

Recall that the Poisson process is a counting process in which the interarrival times are independent and identically distributed exponentially. That is, in a Poisson process, the law of probability of the arrival events (or renewals) obeys the independent exponential distribution with mean $1/\lambda$. To generalize the Poisson process, we can consider a counting process $\{N(t), t \geq 0\}$ in which the interarrival times are independent and distributed exponentially with state dependent mean $1/\lambda_k$, say, for state $k = 0, 1, 2, \cdots$. Such processes arise in arriving particles or customers, population growth, bacteria multiplication, cell division, and radioactive transmutation. In particular, such a process was formulated by applying biological models and is called a pure birth process.

Definition 6.2.1 If a counting process $\{N(t), t \geq 0\}$ is a Markov chain with stationary transition probabilities and satisfies the following:

(i) $N(0) = 0$,

(ii) $P\{N(t+h) - N(t) = 1 \mid N(t) = k\} = \lambda_k h + o(h)$,

(iii) $P\{N(t+h) - N(t) \geq 2 \mid N(t) = k\} = o(h)$,

then the process is called a *pure birth process* with parameters $\{\lambda_k, k = 0, 1, 2, \cdots\}$.

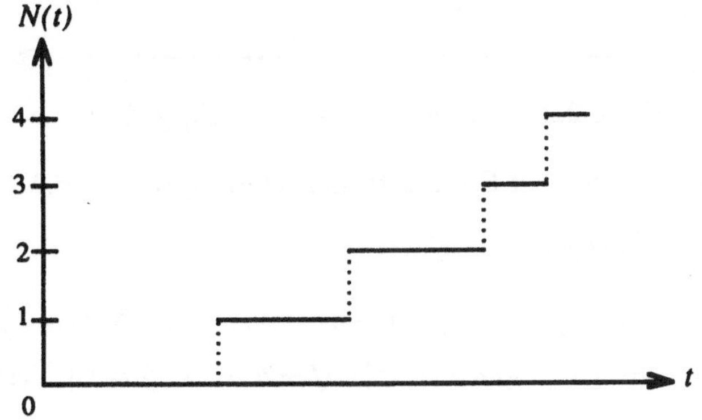

Fig. 6.2.1 A sample function of the pure birth process.

A pure birth process is a counting process and its realization is shown in Fig. 6.2.1. Noting that the process is a Markov chain, we are concerned with the stationary transition probability

$$P_{ij}(t) = P\{N(t) = j \mid N(0) = i\} \quad (i, j = 0, 1, 2, \cdots), \tag{6.2.1}$$

which satisfies the Chapman-Kolmogorov equation in Eq.(6.1.3). The transition probability by specifying the initial condition $N(0) = 0$ in (i) of Definition 6.2.1 is

$$P_k(t) = P\{N(t) = k \mid N(0) = 0\} \quad (k = 0, 1, 2, \cdots) \tag{6.2.2}$$

and can be calculated. We are concerned with the transition probability in Eq.(6.2.2).

Let us consider the probability $P_0(t)$. Specifying the times t and h, where $h > 0$ is the infinitesimal time interval, we have

$$P_0(t+h) = P\{N(t+h) = 0\}$$

$$= P\{N(t) = 0\}P\{N(t+h) - N(t) = 0 \mid N(t) = 0\}$$
$$= P_0(t)[1 - \lambda_0 h + o(h)]. \tag{6.2.3}$$

Rearranging both sides and letting $h \to 0$ imply

$$P_0'(t) = \frac{dP_0(t)}{dt} = \lim_{h\to 0} \frac{P_0(t+h) - P_0(t)}{h} = -\lambda_0 P_0(t). \tag{6.2.4}$$

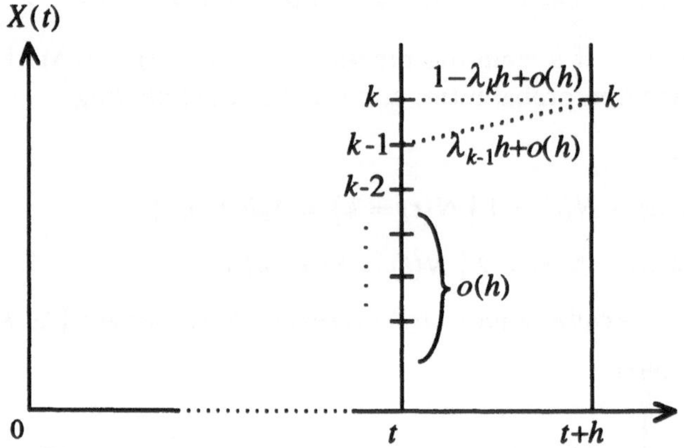

Fig. 6.2.2 An interpretation of Eq.(6.2.6).

Similar to the way we derived Eq.(3.2.9) and referring to Fig. 6.2.2, we have

$$\begin{aligned}P_k(t+h) &= P\{N(t+h) = k\}\\ &= P\{N(t) = k\}P\{N(t+h) - N(t) = 0 \mid N(t) = k\}\\ &\quad +P\{N(t) = k-1\}P\{N(t+h) - N(t) = 1 \mid N(t) = k-1\}\\ &\quad +\sum_{i=2}^{k} P\{N(t) = k-i\}P\{N(t+h) - N(t) = i \mid N(t) = k-i\}\\ &= P_k(t)[1 - \lambda_k h + o(h)] + P_{k-1}(t)[\lambda_{k-1}h + o(h)] + o(h), \end{aligned} \tag{6.2.5}$$

which implies

$$P_k'(t) = \frac{dP_k(t)}{dt} = -\lambda_k P_k(t) + \lambda_{k-1}P_{k-1}(t) \quad (k = 1, 2, \cdots). \tag{6.2.6}$$

A set of the differential equations (or difference-differential equations) (6.2.4) and (6.2.6) is called *Kolmogorov's forward equation* for the pure birth process. Equations (6.2.4) and (6.2.6) can be recursively solved by

$$P_0(t) = e^{-\lambda_0 t}, \tag{6.2.7}$$

$$P_k(t) = \lambda_{k-1}e^{-\lambda_k t}\int_0^t e^{\lambda_k x}P_{k-1}(x)dx \quad (k = 1, 2, \cdots) \tag{6.2.8}$$

under the initial conditions that $P_0(0) = P\{N(0) = 0\} = 1$ and $P_k(0) = P\{N(0) = k\} = 0$ for $k \geq 1$.

As shown in Theorem 3.3.1, the interarrival times are independent and identically distributed exponentially. Since the following theorem just corresponds to Theorem 3.3.1, we will simply state it without proof.

Theorem 6.2.1 For a pure birth process $\{N(t), t \geq 0\}$ with parameters $\{\lambda_k, k = 0, 1, 2, \cdots\}$, the interarrival times $X_{k+1}(k = 0, 1, 2, \cdots)$ are independent and distributed exponentially with parameter λ_k (mean $1/\lambda_k$).

From this theorem, we can understand that each mean interarrival time is $1/\lambda_k$ $(k = 0, 1, \cdots)$. That is, the total mean time to infinity of the events is $\sum_{k=0}^{\infty} 1/\lambda_k$, which should diverge. The following theorem assures that the total probability is a unity for all $t \geq 0$.

Theorem 6.2.2 For a pure birth process $\{N(t), t \geq 0\}$ with parameters $\{\lambda_k, k = 0, 1, 2, \cdots\}$,

$$\sum_{n=0}^{\infty} P_n(t) = 1 \tag{6.2.9}$$

for all $t \geq 0$ if and only if

$$\sum_{k=0}^{\infty} \frac{1}{\lambda_k} = \infty. \tag{6.2.10}$$

The proof needs much elaboration and we omit it. Two simple examples are cited for Theorem 6.2.2.

Example 6.2.1 A Poisson process can be obtained by assuming $\lambda_k = \lambda$ $(k = 0, 1, 2, \cdots)$ for a pure birth process. Of course,

$$\sum_{k=0}^{\infty} \frac{1}{\lambda_k} = \sum_{k=0}^{\infty} \frac{1}{\lambda} = \infty$$

assures Eq.(6.2.9).

Example 6.2.2 Consider a pure birth process with parameter $\lambda_k = 2^k \lambda$. Then

$$\sum_{k=0}^{\infty} \frac{1}{\lambda_k} = \frac{1}{\lambda} \sum_{n=0}^{\infty} \frac{1}{2^n} = \frac{2}{\lambda} < \infty,$$

which implies that the events take place infinitely often within the finite time interval. The process with this example is not valid for treating a pure birth process.

The *Yule process* in biology and the *Furry process* in physics are well-known examples of pure birth processes. Such names for the process were originated by

Yule for his contribution to the mathematical theory of evolution and by Furry for his contribution to a process connected with cosmic rays.

For the Yule process we assume that each member in a population has a probability $\lambda h + o(h)$ of giving birth to a new member in the infinitesimal time interval. Given that there are $N(0) = i$ (i: natural number) members at time 0, and assuming independence and no interaction among members, we have

$$P\{N(t+h) - N(t) = 1 \mid N(t) = k+i\}$$

$$= \binom{k+i}{1} [\lambda h + o(h)][1 - \lambda h + o(h)]^{k+i-1} = (k+i)\lambda h + o(h). \quad (6.2.11)$$

which implies $\lambda_{k+i} = (k+i)\lambda$. Note that Eq.(6.2.10) holds for the Yule process. Note that we have to relabel $k = i, i+1, i+2, \cdots$ for index in Eqs.(6.2.7) and (6.2.8).

Example 6.2.3 (*Yule process*) Assuming $i = 1$ for the Yule process, we can solve $P_k(t)$ $(k = 1, 2, 3, \cdots)$ recursively. That is,

$$P_1(t) = e^{-\lambda t},$$

$$P_2(t) = \lambda e^{-2\lambda t} \int_0^t e^{2\lambda x} P_1(x) dx = e^{-\lambda t}(1 - e^{-\lambda t}),$$

$$P_3(t) = 2\lambda e^{-3\lambda t} \int_0^t e^{3\lambda x} P_2(x) dx = e^{-\lambda t}(1 - e^{-\lambda t})^2,$$

$$\vdots$$

$$P_{k+1}(t) = k\lambda e^{-(k+1)\lambda t} \int_0^t e^{(k+1)\lambda x} P_k(x) dx = e^{-\lambda t}(1 - e^{-\lambda t})^k,$$

where we notice that the transition probability $P_k(t)$ denotes that the process is in state k (k members) at time t, given that it was in state 1 (1 member) at time 0. Naturally, we have to add 1 as an index in Eqs.(6.2.7) and (6.2.8).

The generating function for $P_k(t)$ $(k = 1, 2, \cdots)$ can be easily obtained by summing over k:

$$g(s) = \sum_{k=1}^{\infty} P_k(t) s^k = \frac{se^{-\lambda t}}{1 - (1 - e^{-\lambda t})s} \qquad (0 < \mid s \mid < 1). \quad (6.2.12)$$

Consider the Yule process in which $N(0) = r$ members at time 0 in general. Noting that independence and no interaction among the members is given, we regard the Yule process with $N(0) = r$ as the sum of r independent Yule processes with $N(0) = 1$. Let

$$P_{rn}(t) = P\{N(t) = n \mid X(0) = r\} \quad (n = r, r+1, \cdots) \quad (6.2.13)$$

be the transition probability with $N(0) = r$. Let

$$g_r(s) = \sum_{n=r}^{\infty} P_{rn}(t) s^n \qquad (0 < \mid s \mid < 1) \quad (6.2.14)$$

be the generating function for $P_{rn}(t)$ $(n = r, r+1, \cdots)$. Then we have

$$g_r(s) = [g(s)]^r = \left[\frac{se^{-\lambda t}}{1-(1-e^{-\lambda t})s}\right]^r, \tag{6.2.15}$$

which corresponds to the characteristic function of the negative binomial distribution in Eq.(2.3.25). Assuming $s = e^{iu}, p = e^{-\lambda t}$ in Eq.(2.3.25), we have

$$P_{rn}(t) = \binom{n-1}{n-r} e^{-r\lambda t}(1-e^{-\lambda t})^{n-r}$$

$$= \binom{-r}{n-r} e^{-r\lambda t}(e^{-\lambda t}-1)^{n-r}. \tag{6.2.16}$$

The mean and variance are given by

$$E[N(t)] = re^{\lambda t}, \tag{6.2.17}$$

$$Var[N(t)] = r(1-e^{-\lambda t})e^{2\lambda t}. \tag{6.2.18}$$

6.3 Pure Death Processes

We have developed the pure birth process as one of the generalizations of the Poisson process. We also consider the pure death process.

Definition 6.3.1 If a stochastic process $\{X(t), t \geq 0\}$ is a Markov chain with stationary transition probabilities with state space $\{k, k = 0, 1, 2, \cdots, n\}$, n: a positive integer, and satisfies the following:

(i) $X(0) = n$,

(ii) $P\{X(t+h) - X(t) = -1 \mid X(t) = k\} = \mu_k h + o(h)$,

(iii) $P\{X(t+h) - X(t) \geq -2 \mid X(t) = k\} = o(h)$,

then the process is called a *pure death process* with parameters $\{\mu_k, k = 1, 2, \cdots, n\}$.

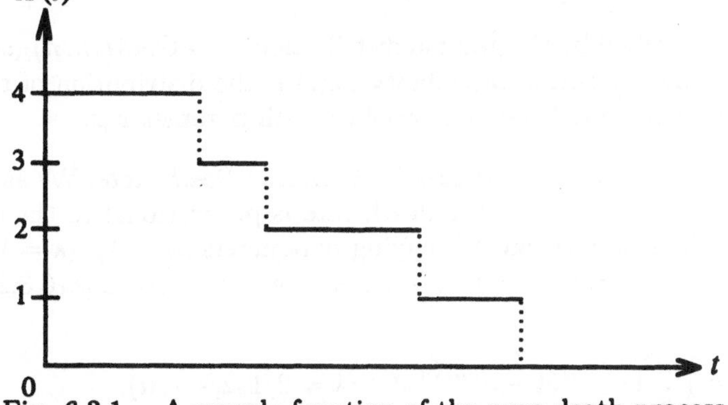

Fig. 6.3.1 A sample function of the pure death process.

It is natural that in the pure birth process, the events take place infinitely often for an infinite interval of time (cf. Theorem 6.2.2). However, in the pure death process, at most n events take place for any interval of time, since no event takes place once the process reaches state 0, i.e., state 0 is an absorbing state. Figure 6.3.1 shows a sample function of the pure death process. Let

$$P_k(t) = P\{X(t) = k \mid X(0) = n\} \quad (k = 0, 1, 2, \cdots, n) \tag{6.3.1}$$

be the transition probability by specifying the initial condition $X(0) = n$ in (i) of Definition 6.3.1. Similar to the derivation of Kolmogorov's forward equation in Eqs.(6.2.4) and (6.2.6), we have

$$P_n'(t) = -\mu_n P_n(t), \tag{6.3.2}$$

$$P_k'(t) = -\mu_k P_k(t) + \mu_{k+1} P_{k+1}(t) \quad (k = 1, 2, \cdots, n-1), \tag{6.3.3}$$

$$P_0'(t) = \mu_1 P_1(t), \tag{6.3.4}$$

which is *Kolmogorov's forward equation* for the pure death process. Specifying parameters $\{\mu_k, k = 1, 2, \cdots, n\}$, we can obtain the transition probability $P_k(t)$ $(k = 0, 1, 2, \cdots, n)$.

Example 6.3.1 Specifying parameters $\mu_k = \mu$ $(k = 1, 2, \cdots, n)$ and $P_n(0) = 1$, $P_k(0) = 0$ $(k = 1, 2, \cdots, n-1)$, we have

$$P_k(t) = \frac{(\mu t)^{n-k}}{(n-k)!} e^{-\mu t} \quad (k = 1, 2, \cdots, n)$$

and

$$\begin{aligned} P_0(t) &= 1 - \sum_{k=1}^{n} P_k(t) \\ &= 1 - \sum_{k=1}^{n} \frac{(\mu t)^{n-k}}{(n-k)!} e^{-\mu t} \\ &= 1 - \sum_{k=0}^{n-1} \frac{(\mu t)^k}{k!} e^{-\mu t}, \end{aligned}$$

which can be described by the gamma distribution $X \sim GAM(\mu, n)$ (see Section 2.4 (iii)), since the transition probability $P_0(t)$ is the distribution of the sum of n independent exponential random variables with parameter μ.

Example 6.3.2 (*Pure Death Process with Linear Death Rate*) We assume that $\mu_k = k\mu$ $(k = 1, 2, \cdots, n)$, i.e., the death rate is proportional to the number of members alive in a population. Specifying parameters $\mu_k = k\mu$ $(k = 1, 2, \cdots, n)$ and $P_n(0) = 1$, $P_k(0) = 0$ $(k = 1, 2, \cdots, n-1)$, we can solve Eqs.(6.3.2) - (6.3.4) whose solution is

$$P_k(t) = \binom{n}{k} e^{-k\mu t}(1 - e^{-\mu t})^{n-k} \quad (k = 0, 1, 2, \cdots, n),$$

which is the binomial distribution $X \sim B(n, e^{-\mu t})$ (see Section 2.3 (iii)). The mean and variance are given by

$$E[X(t)] = ne^{-\mu t},$$

$$Var[X(t)] = ne^{-\mu t}(1 - e^{-\mu t}).$$

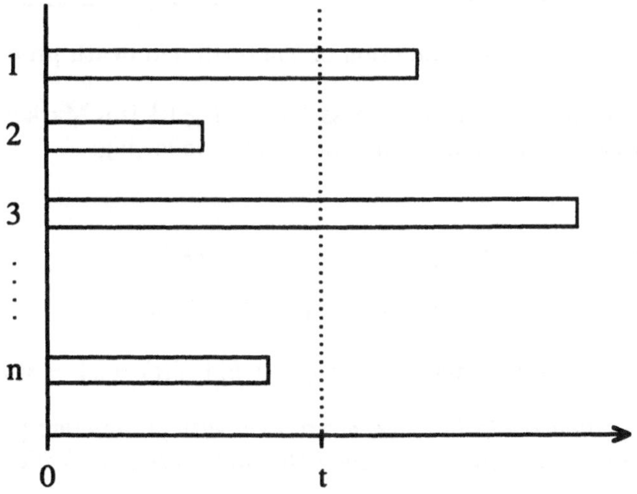

Fig. 6.3.2 An interpretation of the probability $\mathbf{P}_k(t)$ that k members are alive at time t.

The transition probability $P_k(t)$ can be easily derived by considering that k members are alive at time t, given that n members are alive at time 0, where the probability of each member being alive at time t is $e^{-\mu t}$ (see Fig. 6.3.2).

6.4 Birth and Death Processes

We have developed the pure birth process in Section 6.2 and the pure death process in Section 6.3. Combining the pure birth and pure death processes of which a sample function is shown in Fig. 6.4.1, i.e., in a birth and death process, the process moves up and down by yielding a birth and death, respectively. Unless otherwise specified, there is no absorbing state in a birth and death process.

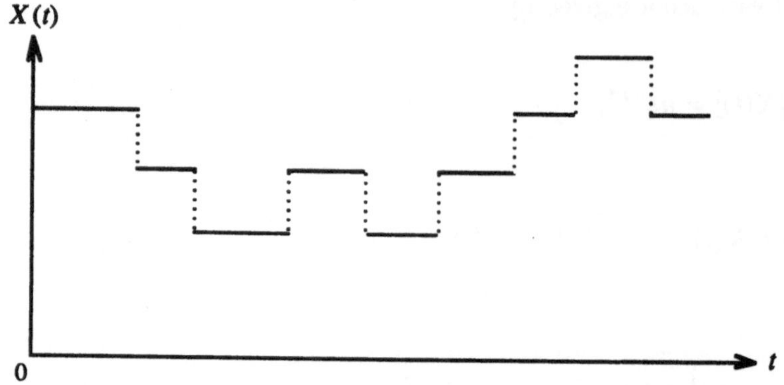

Fig. 6.4.1 A sample function of the birth and death process.

Definition 6.4.1 If a stochastic process $\{X(t),\ t \geq 0\}$ is a Markov chain with stationary transition probabilities and satisfies the following:

(i) $X(0) = i$,

(ii) $P\{X(t+h) - X(t) = 1 \mid X(t) = k\} = \lambda_k h + o(h)$,

(iii) $P\{X(t+h) - X(t) = -1 \mid X(t) = k\} = \mu_k h + o(h)$,

(iv) $P\{\text{two or more events take place in } (t, t+h] \mid X(t) = k\} = o(h)$,

then the process is called a *birth and death process* with parameters $\{\lambda_k, \mu_{k+1}, k = 0, 1, 2, \cdots\}$, where λ_k and μ_{k+1} are called the *birth rate* and *death rate*, respectively.

The conditions in Definition 6.4.1 can be interpreted as follows: (i) gives the initial condition which implies $P_{ii}(0) = 1$ and $P_{ij}(0) = 0$ $(i \neq j)$ (See Eq.(6.4.3)); (ii) and (iii) show that the birth and death rates are λ_k and μ_k, given that the process is in state k; and (iv) shows that the probability that two or more events take place in a small interval is negligible as $h \to 0$. Applying the conditions (ii), (iii), and (iv) of Definition 6.4.1, we have

$$P\{X(t+h) - X(t) = 0 \mid X(t) = 0\} = 1 - \lambda_0 h + o(h), \tag{6.4.1}$$

$$P\{X(t+h) - X(t) = 0 \mid X(t) = k\} = 1 - (\lambda_k + \mu_k)h + o(h) \quad (k = 1, 2, \cdots). \tag{6.4.2}$$

Let

$$P_{ij}(t) = P\{X(t) = j \mid X(0) = i\} \quad (i, j = 0, 1, 2, \cdots) \tag{6.4.3}$$

be the stationary transition probability that the process is in state j at time t, given that it was in state i at time 0. Applying Chapman-Kolmogorov equation in Eq.(6.1.3), and assuming the times t and h, where $h > 0$ is the infinitesimal time interval, we have for $j = 0$,

$$\begin{aligned}
P_{i0}(t+h) &= P_{i0}(t)P\{X(t+h)-X(t)=0 \mid X(t)=0\} \\
&\quad + P_{i1}(t)P\{X(t+h)-X(t)=-1 \mid X(t)=1\} \\
&\quad + \sum_{k=2}^{\infty} P_{ik}(t)P\{X(t+h)-X(t)=-k \mid X(t)=k\} \\
&= P_{i0}(t)[1-\lambda_0 h+o(h)] + P_{i1}(t)[\mu_1 h+o(h)] + o(h). \qquad (6.4.4)
\end{aligned}$$

Rearranging both sides and letting $h \to 0$ imply

$$P'_{i0}(t) = \frac{dP_{i0}(t)}{dt} = -\lambda_0 P_{i0}(t) + \mu_1 P_{i1}(t). \qquad (6.4.5)$$

Similar to deriving Eq.(6.2.5), we have in general for j,

$$\begin{aligned}
P_{ij}(t+h) &= P_{i,j-1}(t)P\{X(t+h)-X(t)=1 \mid X(t)=j-1\} \\
&\quad + P_{ij}(t)P\{X(t+h)-X(t)=0 \mid X(t)=j\} \\
&\quad + P_{i,j+1}(t)P\{X(t+h)-X(t)=-1 \mid X(t)=j+1\} \\
&\quad + \sum_{\substack{k=0 \\ k\neq j-1,j,j+1}}^{\infty} P_{ik}(t)P\{X(t+h)-X(t)=j-k \mid X(t)=k\} \\
&= P_{i,j-1}(t)[\lambda_{j-1}h+o(h)] + P_{ij}(t)[1-(\lambda_j+\mu_j)h+o(h)] \\
&\quad + P_{i,j+1}(t)[\mu_{j+1}h+o(h)] + o(h). \qquad (6.4.6)
\end{aligned}$$

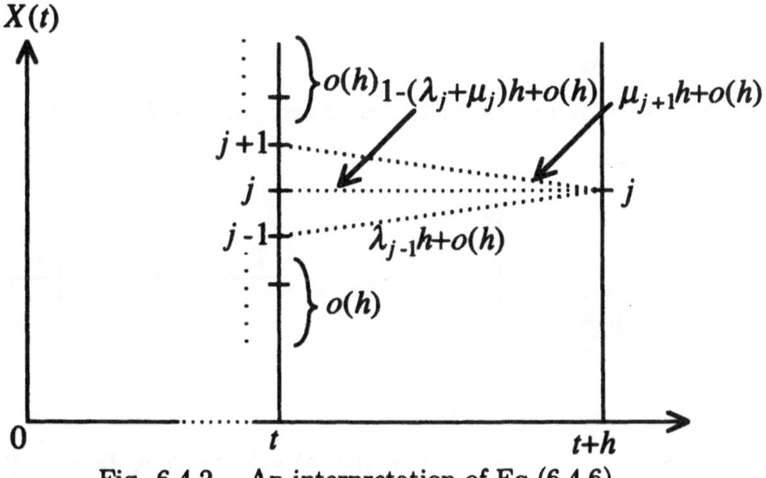

Fig. 6.4.2 An interpretation of Eq.(6.4.6).

Rearranging both sides and letting $h \to 0$ imply

$$P'_{ij}(t) = \lambda_{j-1}P_{i,j-1}(t) - (\lambda_j+\mu_j)P_{ij}(t) + \mu_{j+1}P_{i,j+1}(t)$$

$$(j = 1, 2, \cdots). \tag{6.4.7}$$

The set of differential equations (or difference-differential equations) (6.4.5) and (6.4.7) are called *Kolmogorov's forward equation* for the birth and death process. Applying the initial conditions that

$$P_{ii}(0) = 1, \quad P_{ij}(0) = 0 \quad (i \neq j), \tag{6.4.8}$$

we can solve Eqs.(6.4.5) and (6.4.7) with respect to $P_{ij}(t) \quad (j = 0, 1, 2, \cdots)$ in principle. However, it is difficult or impossible except in the simplest cases to solve them analytically. We will discuss the numerical computation of the transition probabilities for the general finite-state Markov chain. A direct extension of Theorem 6.2.1 is the following, given without proof:

Theorem 6.4.1 For a birth and death process $\{X(t),\ t \geq 0\}$ with parameters $\{\lambda_k,\ \mu_{k+1},\ k = 0, 1, 2, \cdots\}$, when the process is in state i $(i = 0, 1, 2, \cdots)$ at time t, i.e., $X(t) = i$, the interarrival time is distributed exponentially with parameter $\lambda_i + \mu_i$, where the probability of moving to the next state $i - 1$ or $i + 1$ is $\mu_i/(\lambda_i + \mu_i)$ or $\lambda_i/(\lambda_i + \mu_i)$, respectively. Note that when $i = 0$, the possible transition state is only state 1 (i.e., we interpret that $\mu_0 = 0$ implies $\mu_0/(\lambda_0 + \mu_0) = 0$ and $\lambda_0/(\lambda_0 + \mu_0) = 1$).

For the birth and death process $\{X(t),\ t \geq 0\}$, the process has stationary independent increments resulting in the exponential distribution of all interarrival times between transitions. In particular, for a state k, the interarrival time is distributed exponentially with parameter $\lambda_k + \mu_k$, and the process moves to state $k + 1$ with probability $\lambda_k/(\lambda_k + \mu_k)$ and to state $k - 1$ with probability $\mu_k/(\lambda_k + \mu_k)$. Note that

$$\frac{\lambda_k}{\lambda_k + \mu_k} + \frac{\mu_k}{\lambda_k + \mu_k} = 1 \tag{6.4.9}$$

implies that there are no other transitions except states $k - 1$ and $k + 1$.

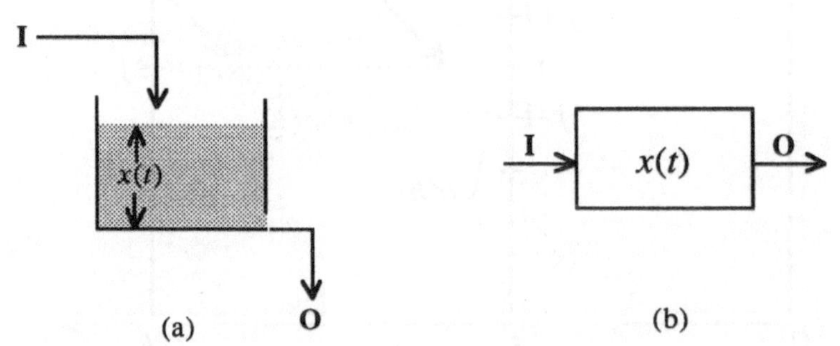

Fig. 6.4.3 A water tank model (a) and its block diagram (b).

To explain the differential equations (6.4.5) and (6.4.7), we consider the analogy of a simple physical model of a *water tank*. There is a water tank whose level is $x(t)$ at time t, where the input amount of water is I per unit time and the output amount of water is O per unit time. The differential equation with respect to level $x(t)$ is given by

$$\frac{dx(t)}{dt} = I - O \tag{6.4.10}$$

(see Fig. 6.4.3 (a)), and a *block diagram* of the water tank model is shown in Fig. 6.4.3 (b). If we consider a two cascade water tank model in Fig. 6.4.4 (a), the corresponding differential equations are given by

$$\frac{dx_1(t)}{dt} = I_1 - O_1, \quad \frac{dx_2(t)}{dt} = I_2 - O_2 \quad (O_1 = I_2), \tag{6.4.11}$$

where $x_1(t)$ and $x_2(t)$ are the levels of tanks 1 and 2 at time t. The corresponding block diagram is shown in Fig. 6.4.4 (b).

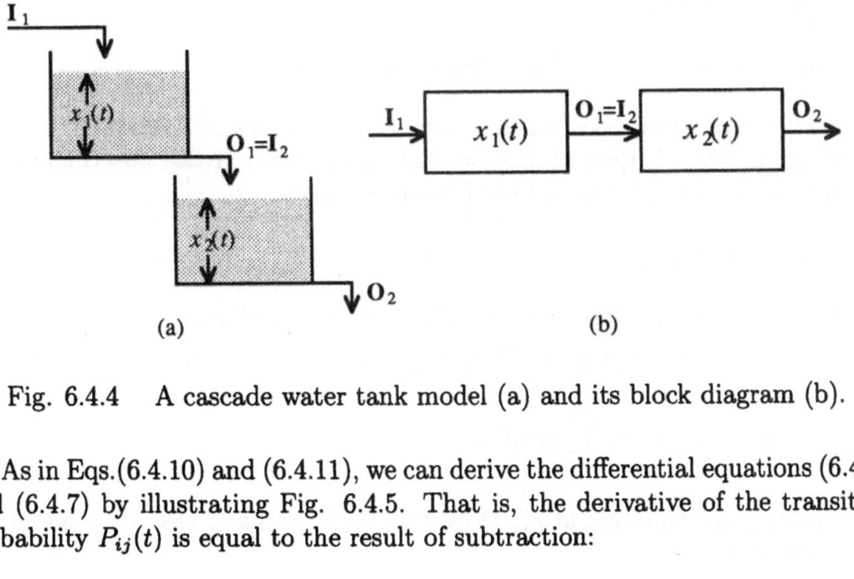

Fig. 6.4.4 A cascade water tank model (a) and its block diagram (b).

As in Eqs.(6.4.10) and (6.4.11), we can derive the differential equations (6.4.5) and (6.4.7) by illustrating Fig. 6.4.5. That is, the derivative of the transition probability $P_{ij}(t)$ is equal to the result of subtraction:

$$\frac{dP_{ij}(t)}{dt} = I_j - O_j$$

$$= \begin{cases} \lambda_{j-1}P_{i,j-1}(t) + \mu_{j+1}P_{i,j+1}(t) - (\lambda_j + \mu_j)P_{ij}(t) & (j = 1, 2, \cdots) \\ \mu_1 P_{i1}(t) - \lambda_0 P_{i0}(t) & (j = 0), \end{cases} \tag{6.4.12}$$

which is Kolmogorov's forward equation for the birth and death process. In particular, the initial condition $P_{ii}(0) = 1$ corresponds to the water tank $P_{ii}(t)$

full of unit water at time $t = 0$. Once the level changes over time, it can be described by Kolmogorov's forward equation.

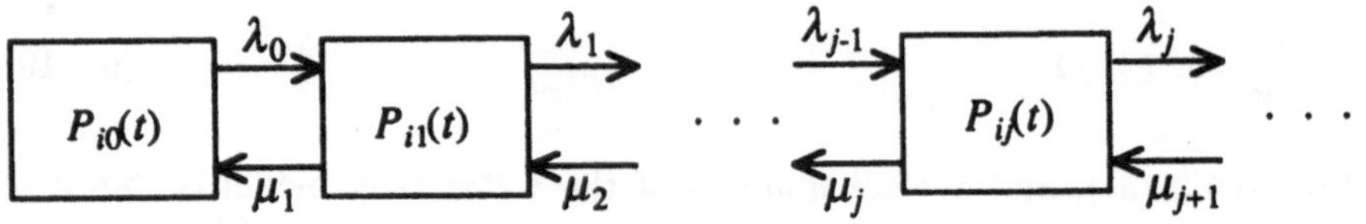

Fig. 6.4.5 A block diagram of the birth and death process.

Let us consider the limiting behavior of the transition probability $P_{ij}(t)$ as $t \to \infty$. As shown above in the water tank model, levels remain constant in the steady-state under suitable conditions. Such behavior will be discussed as the limiting behavior of the Markov chain.

Before discussing this, we give an example of the birth and death processes.

Example 6.4.1 (*Linear Growth Process*) A birth and death process is called a *linear growth process* if

$$\lambda_k = k\lambda, \ \mu_{k+1} = (k+1)\mu \quad (k = 0, 1, 2, \cdots).$$

Examples of such processes arise in the study of biological reproduction and population growth. Noting that $\lambda_0 = 0$ and state 0 is only an absorbing state, we have the following Kolmogorov's forward equation:

$$P'_{i0}(t) = \mu P_{i1}(t),$$

$$P'_{ij}(t) = (j-1)\lambda P_{i,j-1}(t) - j(\lambda+\mu)P_{ij}(t) + (j+1)\mu P_{i,j+1}(t) \quad (j = 1, 2, \cdots).$$

Assuming $X(0) = i \geq 1$, we introduce the expectation at time t:

$$M(t) = E[X(t)] = \sum_{j=0}^{\infty} jP_{ij}(t).$$

If we multiply the equation of $P'_{ij}(t)$ by j on both sides and take the summation over j, we have

$$M'(t) = (\lambda - \mu)M(t)$$

with initial condition $M(0) = i$. The solution of this equation is given by

$$M(t) = ie^{(\lambda-\mu)t}.$$

Let us consider the limiting behavior of $M(t)$ as $t \to \infty$. It is obvious that

$$\lim_{t\to\infty} M(t) = \begin{cases} 0 & (\lambda < \mu) \\ i & (\lambda = \mu) \\ \infty & (\lambda > \mu). \end{cases}$$

That is, if the birth rate λ is greater than the death rate μ, the mean population diverges; if the birth rate is equal to the death rate, the mean population never changes over time; and if the birth rate λ is less than the death rate μ, the population mean converges toward zero, i.e., vanishes. This fact can be easily verified by the differential equation $M'(t) = (\lambda - \mu)M(t)$. We can also discuss the second moment of $X(t)$ by similar techniques using differential equations (see Problem 6.4). In particular, the explicit expressions of $P_{ij}(t)$ can be given by using the techniques of the generating functions (see Problem 6.5).

Let us discuss the limiting behavior of the general birth and death process. We assume in general that all the parameters are positive for the birth and death process with denumerable states $k = 0, 1, 2, \cdots$. That is, we assume that $\lambda_k > 0,\ \mu_{k+1} > 0 \quad (k = 0, 1, 2, \cdots)$. Under the assumption that all the parameters are positive, the process is irreducible and recurrent. However, we have to identify whether the process is positive recurrent or not. Recall that the steady-state probabilities mean that the levels of water never change over time in the water tank model. That is, if there exist the limiting probabilities

$$p_j = \lim_{t\to\infty} P_{ij}(t) \tag{6.4.13}$$

which are independent of the initial state i, then we have

$$-\lambda_0 p_0 + \mu_1 p_1 = 0, \tag{6.4.14}$$

$$\lambda_{j-1}p_{j-1} - (\lambda_j + \mu_j)p_j + \mu_{j+1}p_{j+1} = 0 \quad (j = 1, 2, \cdots), \tag{6.4.15}$$

which are derived by assuming that $P'_{ij}(t) = 0$ and substituting p_j for $P_{ij}(t)$ in Eqs.(6.4.5) and (6.4.7). From the law of total probability, we have

$$\sum_{j=0}^{\infty} p_j = 1. \tag{6.4.16}$$

Let us solve the simultaneous equations in Eqs.(6.4.14)-(6.4.16). Rewriting Eqs.(6.4.14) and (6.4.15), we have

$$\lambda p_0 = \mu_1 p_1, \tag{6.4.17}$$

$$\lambda_1 p_1 - \lambda_0 p_0 = \mu_2 p_2 - \mu_1 p_1, \tag{6.4.18}$$

$$\vdots$$

$$\lambda_{j-1}p_{j-1} - \lambda_{j-2}p_{j-2} = \mu_j p_j - \mu_{j-1}p_{j-1}. \tag{6.4.19}$$

Summing over both sides, we have

$$\lambda_{j-1}p_{j-1} = \mu_j p_j, \tag{6.4.20}$$

which implies

$$p_j = \frac{\lambda_{j-1}}{\mu_j}p_{j-1} = \frac{\lambda_{j-1}\lambda_{j-2}}{\mu_j\mu_{j-1}}p_{j-2} = \cdots = \frac{\lambda_{j-1}\lambda_{j-2}\cdots\lambda_0}{\mu_j\mu_{j-1}\cdots\mu_1}p_0 = \left(\prod_{k=1}^{j}\frac{\lambda_{k-1}}{\mu_k}\right)p_0.$$

(6.4.21)

Using the law of total probability in Eq.(6.4.16), we have

$$\left[1 + \sum_{j=1}^{\infty} \prod_{k=1}^{j} \frac{\lambda_{k-1}}{\mu_k}\right] p_0 = 1. \tag{6.4.22}$$

The existence of the limiting probabilities $p_j > 0$ $(j = 0, 1, 2, \cdots)$ presupposes the following necessary condition:

$$1 + \sum_{j=1}^{\infty} \prod_{k=1}^{j} \frac{\lambda_{k-1}}{\mu_k} = \sum_{j=0}^{\infty} \prod_{k=1}^{j} \frac{\lambda_{k-1}}{\mu_k} < \infty, \tag{6.4.23}$$

where we postulate $\prod_{k=1}^{j} \cdot = 1$ for $j = 0$ for simplicity of notation. It is well-known that the necessary condition of Eq.(6.4.23) is also the sufficient condition for the existence of the limiting probabilities $p_j > 0$ $(j = 0, 1, 2, \cdots)$. Of course, if the left-hand side of Eq.(6.4.23) diverges, then there exist no limiting probabilities p_j. Summarizing the facts above, we have the following theorem:

Theorem 6.4.2 For a birth and death process with parameters $\{\lambda_k, \mu_{k+1}, k = 0, 1, 2, \cdots\}$, if we assume that all the parameters are positive, i.e.,

$$\lambda_k > 0, \ \mu_{k+1} > 0 \quad (k = 0, 1, 2, \cdots), \tag{6.4.24}$$

there exist the limiting probabilities

$$p_j = \lim_{t \to 0} P_{ij}(t) \quad (i, j = 0, 1, 2, \cdots) \tag{6.4.25}$$

which are independent of the initial state i if and only if

$$\sum_{j=0}^{\infty} \prod_{k=1}^{j} \frac{\lambda_{k-1}}{\mu_k} < \infty, \tag{6.4.26}$$

where we postulate $\prod_{k=1}^{j} \cdot = 1$ for $j = 0$. Then the limiting probabilities are given by

$$p_0 = \left[\sum_{j=0}^{\infty} \prod_{k=1}^{j} \frac{\lambda_{k-1}}{\mu_k}\right]^{-1}, \tag{6.4.27}$$

$$p_j = \left(\prod_{k=1}^{j} \frac{\lambda_{k-1}}{\mu_k}\right) p_0. \tag{6.4.28}$$

Example 6.4.2 (*M/M/1 Queue*) As an example of the birth and death process, we cite an $M/M/1$ queue, where potential customers arrive at Poisson rate λ

and are served exponentially at rate μ by a single channel, and the queue size is infinity. The details will be discussed in Chapter 9. Then $\lambda_k = \lambda$, and $\mu_{k+1} = \mu$ $(k = 0, 1, 2, \cdots)$ for a birth and death process. Verifying the necessary and sufficient condition in Eq.(6.4.26), we have

$$\sum_{j=0}^{\infty} \prod_{k=1}^{j} \left(\frac{\lambda}{\mu}\right) = \sum_{j=0}^{\infty} \left(\frac{\lambda}{\mu}\right)^j = \frac{1}{1-\rho} < \infty$$

if and only if $\rho < 1$, where $\rho = \lambda/\mu$ is called the *traffic intensity* of the system. That is, if the arrival rate (i.e., birth rate) λ is less than the service rate (i.e., death rate) μ, there exist the limiting probabilities

$$p_j = (1-\rho)\rho^j \quad (\rho < 1; j = 0, 1, 2, \cdots),$$

which is the geometric distribution $X \sim GEO(1-\rho)$ in Section 2.3 (iv).

Example 6.4.3 ($M/M/\infty$ *Queue*) As another example of the birth and death process, we consider an $M/M/\infty$ queue, where potential customers arrive at Poisson rate λ and are served exponentially at rate μ by an infinite number of channels (i.e., all the arriving customers are served immediately). Then $\lambda_k = \lambda$, $\mu_{k+1} = (k+1)\mu$ $(k = 0, 1, 2, \cdots)$ for a birth and death process. Verifying the condition in Eq.(6.4.26) and putting $u = \lambda/\mu$, we have

$$\sum_{j=0}^{\infty} \prod_{k=1}^{j} \frac{\lambda_{k-1}}{\mu_k} = \sum_{j=0}^{\infty} \frac{1}{j!}\left(\frac{\lambda}{\mu}\right)^j = e^u < \infty,$$

which is valid for any traffic intensity u. That is, for any λ and μ, there exist the limiting probabilities

$$p_j = \frac{u^j}{j!}e^{-u} \quad (j = 0, 1, 2, \cdots),$$

which is the Poisson distribution $X \sim POI(u)$.

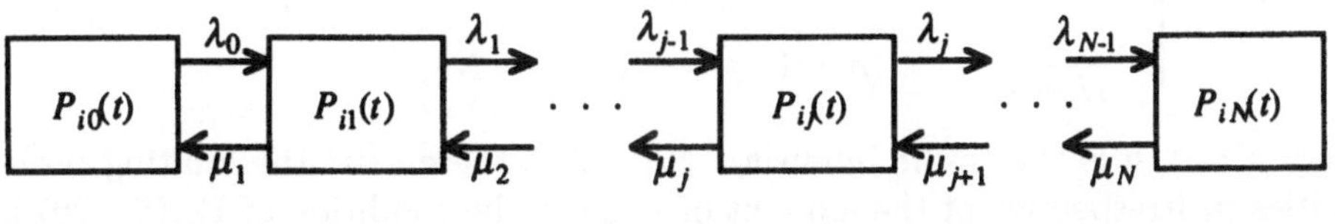

Fig. 6.4.6 A block diagram of the finite-state birth and death process.

We are also interested in a birth and death process with finite state space. Let the state space be $i = 0, 1, 2, \cdots, N$, where N is finite. Then we show a block diagram describing Kolmogorov's forward equation in Fig. 6.4.6. As in Eqs.(6.4.5) and (6.4.7), we have the following Kolmogorov's forward equations:

$$P'_{i0}(t) = -\lambda_0 P_{i0}(t) + \mu_1 P_{i1}(t), \tag{6.4.29}$$

$$P'_{ij}(t) = \lambda_{j-1}P_{i,j-1}(t) - (\lambda_j + \mu_j)P_{ij}(t) + \mu_{j+1}P_{i,j+1}(t) \quad (j = 1, 2, \cdots, N-1), \tag{6.4.30}$$

$$P'_{iN}(t) = \lambda_{N-1}P_{i,N-1}(t) - \mu_N P_{iN}(t). \tag{6.4.31}$$

We are now in a position to show the following:

Theorem 6.4.3 For a finite-state birth and death process with parameters $\{\lambda_k,\ \mu_{k+1},\ k = 0, 1, 2, \cdots, N\}$, where N is finite, if we assume that all parameters are positive, i.e.,

$$\lambda_k > 0,\ \mu_{k+1} > 0 \quad (k = 0, 1, 2, \cdots, N-1), \tag{6.4.32}$$

then there exist the limiting probabilities

$$p_j = \lim_{t\to\infty} P_{ij}(t) = \begin{cases} \left[\sum_{j=0}^{N} \prod_{k=1}^{j} \frac{\lambda_{k-1}}{\mu_k}\right]^{-1} & (j = 0) \\ \left(\prod_{k=1}^{j} \frac{\lambda_{k-1}}{\mu_k}\right) p_0 & (j = 1, 2, \cdots, N), \end{cases} \tag{6.4.33}$$

which are independent of the initial state i, where we postulate $\prod_{k=1}^{j} \cdot = 1$ for $j = 0$.

Example 6.4.4 (*$M/M/1/N$ Queue*) Consider an $M/M/1/N$ queue, where potential customers arrive at Poisson rate λ and are served exponentially at rate μ by a single channel, where the maximum system size (including a customer served) is $N < \infty$. Applying Theorem 6.4.3, we have

$$p_j = \begin{cases} \frac{(1-\rho)\rho^j}{1-\rho^{N+1}} & (\rho \neq 1;\ j = 0, 1, 2, \cdots, N) \\ \frac{1}{N+1} & (\rho = 1;\ j = 0, 1, 2, \cdots, N), \end{cases}$$

where $\rho = \lambda/\mu$ is the traffic intensity. Note that there exist the limiting probabilities p_j irrespective of the amount of ρ, since the condition of Eq.(6.4.26) is always satisfied for the finite-state Markov chain.

Example 6.4.5 (*Two-state Markov chain*) For a two-state Markov chain, we have the following Kolmogorov's forward equations:

$$\begin{cases} P'_{i0}(t) = -\lambda P_{i0}(t) + \mu P_{i1}(t) & (i = 0, 1), \\ P'_{i1}(t) = \lambda P_{i0}(t) - \mu P_{i1}(t) & (i = 0, 1), \end{cases}$$

where we assume $\lambda_0 = \lambda$ and $\mu_1 = \mu$, for simplicity. A block diagram is shown in Fig. 6.4.7. Applying Theorem 6.4.3, we have the limiting probabilities:

$$p_0 = \left[1 + \frac{\lambda}{\mu}\right]^{-1} = \frac{\mu}{\lambda+\mu}, \qquad p_1 = \left(\frac{\lambda}{\mu}\right)p_0 = \frac{\lambda}{\lambda+\mu}.$$

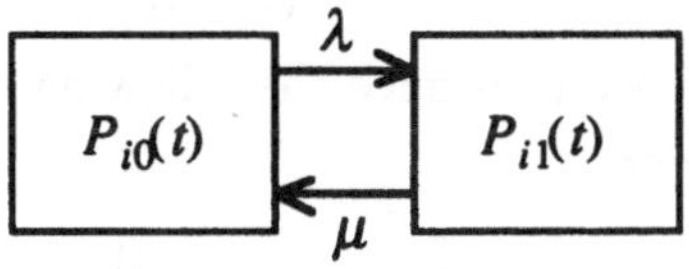

Fig. 6.4.7 A block diagram of the two-state birth and death process.

Solving the differential equations above under the initial conditions that $P_{00}(0) = 1$ and $P_{01}(0) = 0$, we have

$$P_{00}(t) = \frac{\mu}{\lambda+\mu} + \frac{\lambda}{\lambda+\mu}e^{-(\lambda+\mu)t},$$

$$P_{01}(t) = \frac{\lambda}{\lambda+\mu} - \frac{\lambda}{\lambda+\mu}e^{-(\lambda+\mu)t}$$

(see Example 6.5.2). We have the probabilities under the initial conditions that $P_{10}(0) = 0,\ P_{11}(0) = 1$:

$$P_{10}(t) = \frac{\mu}{\lambda+\mu} - \frac{\mu}{\lambda+\mu}e^{-(\lambda+\mu)t},$$

$$P_{11}(t) = \frac{\lambda}{\lambda+\mu} + \frac{\mu}{\lambda+\mu}e^{-(\lambda+\mu)t}.$$

In any case, we have the limiting probabilities p_j $(j = 0, 1)$ which are independent of the initial distribution.

Let us assume the following initial distribution:

$$P\{X(0) = 0\} = p_0 = \frac{\mu}{\lambda+\mu}, \quad P\{X(0) = 1\} = p_1 = \frac{\lambda}{\lambda+\mu}.$$

Then the transition probability distribution at time t under the initial distribution above is given by

$$\begin{bmatrix} \frac{\mu}{\lambda+\mu} & \frac{\lambda}{\lambda+\mu} \end{bmatrix} \begin{bmatrix} P_{00}(t) & P_{01}(t) \\ P_{10}(t) & P_{11}(t) \end{bmatrix} = \begin{bmatrix} \frac{\mu}{\lambda+\mu} & \frac{\lambda}{\lambda+\mu} \end{bmatrix}$$

which is independent of the time t and is just the same as the limiting probabilities (see Problem 6.6).

We have derived the Kolmogorov's forward equations in Eqs.(6.4.5) and (6.4.7) by assuming the times t and h, and letting $h \to 0$ for the Chapman-Kolmogorov equation in Eq.(6.1.3). Assuming the times h and t, and letting

$h \to 0$ in Eq.(6.1.3) for the birth and death process, we have *Kolmogorov's backward equations*:

$$P'_{0j}(t) = -\lambda_0 P_{0j}(t) + \lambda_0 P_{1j}(t) \qquad (j = 0, 1, 2, \cdots), \quad (6.4.34)$$

$$P'_{ij}(t) = \mu_i P_{i-1,j}(t) - (\lambda_i + \mu_i) P_{ij}(t) + \lambda_i P_{i+1,j}(t)$$
$$(i = 1, 2, \cdots, j = 0, 1, 2, \cdots). \quad (6.4.35)$$

To express a matrix form, we introduce the following *infinitesimal generator* for the birth and death process:

$$\mathbf{A} = \begin{bmatrix} -\lambda_0 & \lambda_0 & 0 & 0 & \cdots \\ \mu_1 & -(\lambda_1 + \mu_1) & \lambda_1 & 0 & \cdots \\ 0 & \mu_2 & -(\lambda_2 + \mu_2) & \lambda_2 & \cdots \\ 0 & 0 & \mu_3 & -(\lambda_3 + \mu_3) & \cdots \\ \vdots & \vdots & \vdots & \vdots & \ddots \end{bmatrix}. \quad (6.4.36)$$

Introducing the transition probability matrix $\mathbf{P}(t) = [P_{ij}(t)]$, we can express Kolmogorov's forward equations in Eqs.(6.4.5) and (6.4.7) in matrix form:

$$\mathbf{P}'(t) = \mathbf{P}(t)\mathbf{A}, \quad (6.4.37)$$

where $\mathbf{P}'(t) = [P'_{ij}(t)]$. Note that the initial condition is given by

$$\mathbf{P}(0) = \mathbf{I}, \quad (6.4.38)$$

where $\mathbf{I}$ is an identity matrix.

Kolmogorov's backward equations in Eqs.(6.4.34) and (6.4.35) can be expressed in matrix form:

$$\mathbf{P}'(t) = \mathbf{A}\mathbf{P}(t), \quad (6.4.39)$$

where again the initial condition is given in Eq.(6.4.38).

If both the forward equation in Eq.(6.4.37) and the backward equation in Eq.(6.4.39) have unique solutions, respectively, then the solutions are identical and are given by

$$\mathbf{P}(t) = e^{\mathbf{A}t} = \mathbf{I} + \sum_{n=1}^{\infty} \frac{\mathbf{A}^n t^n}{n!}. \quad (6.4.40)$$

Throughout this book, the solutions for the forward and backward equations are identical. However, distinct solutions exist for both equations if we consider the Markov chain with nonstationary probabilities.

Example 6.4.6 Consider a Poisson process with parameter λ. If we assume that $N(0) = i$, i.e., the initial state is in state i at time 0, then

$$P_{ij}(t) = 0 \quad (j < i), \qquad P_{ij}(t) = \frac{(\lambda t)^{j-i}}{(j-i)!} e^{-\lambda t} \quad (j \geq i).$$

Kolmogorov's forward equation in Eq.(3.2.9) is given by

$$P'_{ij}(t) = -\lambda P_{ij}(t) + \lambda P_{i,j-1}(t).$$

We can then note that $P_{i+1,j}(t) = P_{i,j-1}(t)$ since we recognize the stationary transition probability that $(j-1)-i = j-(i+1)$ events take place for the interval of time t. That is,

$$P'_{ij}(t) = -\lambda P_{ij}(t) + \lambda P_{i+1,j}(t)$$

which is Kolmogorov's backward equation for the Poisson process.

6.5 Finite-State Markov Chains

In the preceding sections, we have discussed the pure birth process, pure death process and the birth and death process in which the process can only move to the nearest neighbor state. In general, the process can move to any state, this representing a general Markov chain. To simplify the discussion, we restrict ourselves to finite-state Markov chains in this section. Let us cite the following example of a continuous-time finite-state Markov chain.

Example 6.5.1 Consider an item such as a machine, computer, or other device. The law of failure of the item obeys the exponential distribution with parameter λ. Upon failure, we have to repair the item in two consecutive phases, I and II. The laws of repair of the failed item obey the exponential distributions with parameters μ_1 and μ_2 for phases I and II, respectively. Let state 0 be the state in which the process is working, states 1 and 2, when it is in the two repair phases. Let us consider the transition probability:

$$P_{ij}(t) = P\{X(t) = j \mid X(0) = i\} \quad (i,j = 0,1,2).$$

Applying the Chapman-Kolmogorov equation in Eq.(6.1.3), we have

$$P_{00}(t+h) = P_{00}(t)[1-\lambda h + o(h)] + P_{02}(t)[\mu_2 h + o(h)] + o(h),$$

which implies

$$P'_{00}(t) = -\lambda P_{00}(t) + \mu_2 P_{02}(t).$$

In the same way that we derived $P'_{00}(t)$, we have

$$P'_{01}(t) = -\mu_1 P_{01}(t) + \lambda P_{00}(t),$$

$$P'_{02}(t) = -\mu_2 P_{02}(t) + \mu_1 P_{01}(t),$$

(see Fig. 6.5.1). A set of differential equations can be solved by assuming the initial state (e.g., $X(0) = 0$). It is quite easy to discuss the limiting probabilities

$$p_j = \lim_{t\to\infty} P_{ij}(t) \quad (i,j = 0,1,2).$$

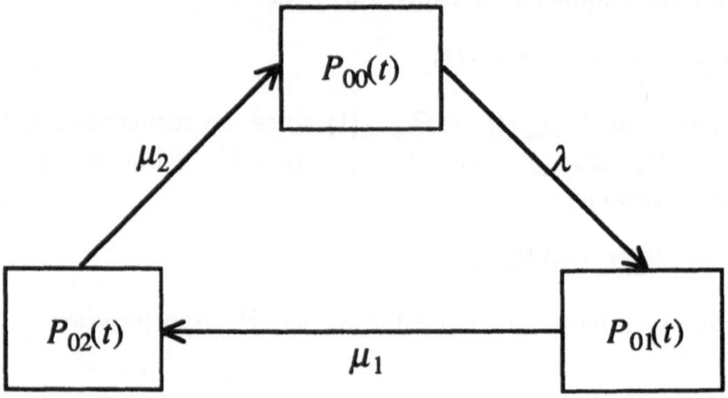

Fig. 6.5.1 A block diagram of the job shop model.

Just as we derived Eqs.(6.4.14) - (6.4.16), we have

$$-\lambda p_0 + \mu_2 p_2 = 0,$$

$$-\mu_1 p_1 + \lambda p_0 = 0,$$

$$-\mu_2 p_2 + \mu_1 p_1 = 0,$$

$$p_0 + p_1 + p_2 = 1.$$

Solving the simultaneous equations above, we have

$$p_0 = \left(1 + \frac{\lambda}{\mu_1} + \frac{\lambda}{\mu_2}\right)^{-1}, \quad p_1 = \left(\frac{\lambda}{\mu_1}\right) p_0, \quad p_2 = \left(\frac{\lambda}{\mu_2}\right) p_0.$$

We do not need to use all the simultaneous equations. It is obvious that the limiting probabilities exist.

In general, we consider the stationary transition probability

$$P_{ij}(t) = P\{X(t) = j \mid X(0) = i\} \quad (i, j = 0, 1, 2, \cdots, N),$$

for the continuous-time finite-state Markov chain $\{X(t),\ t \geq 0\}$ with the state space $\{0, 1, 2 \cdots, N\}$, where N is finite. From the Markov property we have

(i) $P_{ij}(t) \geq 0,$ (6.5.1)

(ii) $\sum_{j=0}^{N} P_{ij}(t) = 1 \quad (i, j = 0, 1, 2, \cdots, N),$ (6.5.2)

(iii) $P_{ij}(t+s) = \sum_{k=0}^{N} P_{ik}(t) P_{kj}(s) \quad (i, j = 0, 1, 2, \cdots, N),$ (6.5.3)

(iv) $\lim_{t \to 0} P_{ij}(t) = \begin{cases} 1 & (i = j) \\ 0 & (i \neq j). \end{cases}$ (6.5.4)

The Chapman-Kolmogorov equation in (iii) above can be written in the matrix form:

$$\mathbf{P}(t+s) = \mathbf{P}(t)\mathbf{P}(s),$$

where $\mathbf{P}(t) = [P_{ij}(t)]_{i,j=0}^{N}$ and $\mathbf{P}(0) = \mathbf{I}$ (an identity matrix). We assume that for an infinitesimal interval of time $h > 0$,

$$P_{ij}(h) = a_{ij}h + o(h) \quad (i \neq j), \tag{6.5.5}$$

where a_{ij} is the transition rate from state i to state j, and is constant. Recall that λ_k and μ_{k+1} are the transition rates for the birth and death process. If the process can move from state i to state j, then $a_{ij} > 0$. Otherwise, $a_{ij} = 0$. Rewriting Eq.(6.5.5), we have

$$\lim_{h\to 0} \frac{P_{ij}(h)}{h} = a_{ij} \quad (i \neq j). \tag{6.5.6}$$

For $h > 0$, we have the law of total probability:

$$P_{ii}(h) + \sum_{\substack{j=0\\ j\neq i}}^{N} P_{ij}(h) = 1, \tag{6.5.7}$$

which implies

$$1 - P_{ii}(h) = \sum_{\substack{j=0\\ j\neq i}}^{N} P_{ij}(h) = \sum_{\substack{j=0\\ j\neq i}}^{N} a_{ij}h + o(h). \tag{6.5.8}$$

Let us define

$$a_i = \lim_{h\to 0} \frac{1 - P_{ii}(h)}{h} = \sum_{\substack{j=0\\ j\neq i}}^{N} a_{ij}, \tag{6.5.9}$$

and introduce the *infinitesimal generator* $\mathbf{A}$ with orthogonal element $-a_i$ and nonorthogonal element a_{ij} $(i \neq j)$:

$$\mathbf{A} = \begin{bmatrix} -a_0 & a_{01} & \cdots & a_{0N} \\ a_{10} & -a_1 & \cdots & a_{1N} \\ \vdots & \vdots & \ddots & \vdots \\ a_{N0} & a_{N1} & \cdots & -a_N \end{bmatrix}. \tag{6.5.10}$$

Rewriting Eqs.(6.5.6) and (6.5.9) in matrix form, we have

$$\lim_{h\to 0} \frac{\mathbf{P}(h) - \mathbf{I}}{h} = \mathbf{A}. \tag{6.5.11}$$

Applying the Chapman-Kolmogorov equation to the matrix form, we have

$$\lim_{h\to 0} \frac{\mathbf{P}(t)\mathbf{P}(h) - \mathbf{P}(t)}{h} = \lim_{h\to 0} \frac{\mathbf{P}(t+h) - \mathbf{P}(t)}{h} = \mathbf{P}(t)\mathbf{A}. \tag{6.5.12}$$

which implies

$$\mathbf{P}'(t) = \mathbf{P}(t)\mathbf{A} \tag{6.5.13}$$

which is *Kolmogorov's forward equation* for the finite-state Markov chain.

Applying Chapman-Kolmogorov equation $\mathbf{P}(t+h) = \mathbf{P}(h)\mathbf{P}(t)$, we have

$$\mathbf{P}'(t) = \mathbf{A}\mathbf{P}(t) \tag{6.5.14}$$

which is *Kolmogorov's backward equation* for the finite-state Markov chain (cf. Fig. 6.5.2).

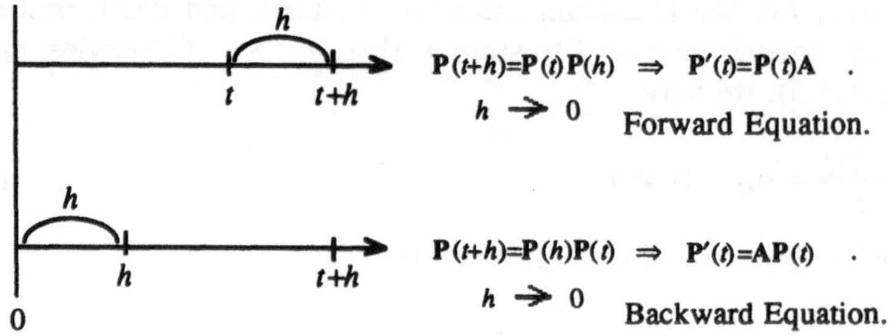

Fig. 6.5.2 An interpretation of Kolmogorov's forward and backward equations.

Under the initial condition $\mathbf{P}(0) = \mathbf{I}$, we can solve Eqs.(6.5.13) and (6.5.14):

$$\mathbf{P}(t) = e^{\mathbf{A}t} = \mathbf{I} + \sum_{n=1}^{\infty} \frac{\mathbf{A}^n t^n}{n!} \tag{6.5.15}$$

which is a unique solution for both forward and backward equations in Eqs.(6.5.13) and (6.5.14).

In principle, if we can calculate all the eigenvalues and their associated column eigenvectors for infinitesimal generator $\mathbf{A}$, we can calculate $\mathbf{P}(t)$ analytically. That is, using all the eigenvalues $\omega_0, \omega_1, \cdots, \omega_N$ of $\mathbf{A}$ and their associated right eigenvectors $\mathbf{u}_0, \mathbf{u}_1, \cdots, \mathbf{u}_N$, if we assume that all the eigenvalues are different, we have

$$\mathbf{A} = \mathbf{U}\mathbf{\Lambda}\mathbf{U}^{-1} \tag{6.5.16}$$

where $\mathbf{U} = [\mathbf{u}_0, \mathbf{u}_1, \cdots, \mathbf{u}_N]$, $\mathbf{U}^{-1}$ is the inverse matrix, and $\mathbf{\Lambda}$ is the orthogonal matrix. Noting that

$$\begin{aligned}\mathbf{A}^n &= \overbrace{(\mathbf{U}\mathbf{\Lambda}\mathbf{U}^{-1})(\mathbf{U}\mathbf{\Lambda}\mathbf{U}^{-1})\cdots(\mathbf{U}\mathbf{\Lambda}\mathbf{U}^{-1})}^{n} \\ &= \mathbf{U}\mathbf{\Lambda}^n\mathbf{U}^{-1},\end{aligned} \tag{6.5.17}$$

we have

$$\mathbf{P}(t) = \mathbf{I} + \sum_{n=1}^{\infty} \frac{\mathbf{U}\mathbf{\Lambda}^n\mathbf{U}^{-1}t^n}{n!} = \mathbf{U}e^{\mathbf{\Lambda}t}\mathbf{U}^{-1}. \tag{6.5.18}$$

Example 6.5.2 (*Example 6.4.5*) Consider a two-state birth and death process whose infinitesimal generator is

$$\mathbf{A} = \begin{bmatrix} -\lambda & \lambda \\ \mu & -\mu \end{bmatrix}.$$

Two eigenvalues are $\omega_0 = 0$ and $\omega_1 = -(\lambda + \mu)$, and we have

$$\mathbf{A} = \mathbf{U\Lambda U}^{-1} = \begin{bmatrix} 1 & \lambda \\ 1 & -\mu \end{bmatrix} \begin{bmatrix} 0 & 0 \\ 0 & -(\lambda+\mu) \end{bmatrix} \begin{bmatrix} \frac{\mu}{\lambda+\mu} & \frac{\lambda}{\lambda+\mu} \\ \frac{1}{\lambda+\mu} & -\frac{1}{\lambda+\mu} \end{bmatrix},$$

which implies

$$\mathbf{P}(t) = \mathbf{U}e^{\mathbf{\Lambda}t}\mathbf{U}^{-1} = \begin{bmatrix} 1 & \lambda \\ 1 & -\mu \end{bmatrix} \begin{bmatrix} 1 & 0 \\ 0 & e^{-(\lambda+\mu)t} \end{bmatrix} \begin{bmatrix} \frac{\mu}{\lambda+\mu} & \frac{\lambda}{\lambda+\mu} \\ \frac{1}{\lambda+\mu} & -\frac{1}{\lambda+\mu} \end{bmatrix}$$

$$= \frac{1}{\lambda+\mu} \begin{bmatrix} \mu + \lambda e^{-(\lambda+\mu)t} & \lambda - \lambda e^{-(\lambda+\mu)t} \\ \mu - \mu e^{-(\lambda+\mu)t} & \lambda + \mu e^{-(\lambda+\mu)t} \end{bmatrix},$$

which has been given in Example 6.4.5.

As shown above, we can compute the transition probability matrix $\mathbf{P}(t) = e^{\mathbf{A}t}$ in principle. However, it is difficult or impossible to do so except in the simplest cases.

The following method is one of the computational methods of $e^{\mathbf{A}t}$, which is called *randomization* or *uniformization*. Let $\mathbf{Q}$ be the transform matrix:

$$\mathbf{Q} = \mathbf{A}/\Lambda + \mathbf{I} \tag{6.5.19}$$

from $\mathbf{A}$, where say, $\Lambda = \max_i a_i$. Note that $\mathbf{Q}$ is the transition probability matrix for the discrete-time Markov chain since each row sum of $\mathbf{Q}$ is a unity and all the elements are non-negative. Note also that the structure of $\mathbf{A}$ is preserved for the transformed Markov chain $\mathbf{Q}$ (i.e., each state classification for $\mathbf{A}$ is preserved for $\mathbf{Q}$). Substituting $\mathbf{A} = \Lambda\mathbf{Q} - \Lambda\mathbf{I}$ (from Eq.(6.5.19)) into $\mathbf{P}(t) = e^{\mathbf{A}t}$, we have

$$\mathbf{P}(t) = e^{\Lambda t\mathbf{Q}}e^{-\Lambda t\mathbf{I}} = e^{-\Lambda t}e^{\Lambda t\mathbf{Q}} = \left[\sum_{n=0}^{\infty} q_{ij}^n \frac{(\Lambda t)^n}{n!} e^{-\Lambda t}\right], \tag{6.5.20}$$

where $\mathbf{Q}^n = [q_{ij}^n]$, an n-step transition probability matrix. The right-hand side of Eq.(6.5.20) is an infinite series of an n-step transition probability by means of the Poisson probability mass function $\frac{(\Lambda t)^n}{n!}e^{-\Lambda t}$. Noting that $\mathbf{Q}^n$ converges toward the stationary probabilities for recurrent states and the unimodal property of the Poisson probability mass functions, we can compute Eq.(6.5.20) for finite terms within a prespecified allowed error, say ε, instead of for an infinite series.

As shown in Theorems 6.2.1 and 6.4.1, the interarrival times are independent and distributed exponentially. We can also show the following theorem for the general Markov chain.

Theorem 6.5.1 For a finite-state Markov chain whose transitions are described by the infinitesimal generator **A**, when the process is in state i ($i = 0, 1, 2, \cdots, N$), the interarrival time to the next transition is independent of all others and is distributed exponentially with parameter a_i, where the probability of moving to the next state j ($j \neq i$) is a_{ij}/a_i.

It is possible to classify each state of the Markov chain by noting that $i \to j$ is equivalent to $a_{ij} > 0$. For the following theorem, we assume that all states communicate with each other, i.e., the Markov chain is irreducible and positive recurrent (since the state space is finite). Note that the Markov chain is aperiodic.

Theorem 6.5.2 If all states communicate with each other for a finite-state Markov chain, there exist the limiting probabilities

$$\lim_{t\to\infty} P_{ij}(t) = p_j > 0 \qquad (j = 0, 1, 2, \cdots, N), \tag{6.5.21}$$

which are independent of the initial state i. Let

$$\mathbf{p} = [p_0 \quad p_1 \quad \cdots \quad p_N] \tag{6.5.22}$$

be the stationary distribution vector of the Markov chain. Then **p** is a unique and positive solution to

$$\mathbf{pA} = \mathbf{0} \tag{6.5.23}$$

$$\sum_{j=0}^{N} p_j = 1. \tag{6.5.24}$$

Example 6.5.3 (*Example 6.4.5*) Consider a two-state birth and death process whose infinitesimal generator is

$$\mathbf{A} = \begin{bmatrix} -\lambda & \lambda \\ \mu & -\mu \end{bmatrix}.$$

From Eqs.(6.5.23) and (6.5.24), we have

$$-\lambda p_0 + \mu p_1 = 0,$$

$$\lambda p_0 - \mu p_1 = 0,$$

$$p_0 + p_1 = 0.$$

Solving the simultaneous equations above, we have

$$\mathbf{p} = [p_0 \quad p_1] = \left[\frac{\mu}{\lambda+\mu} \quad \frac{\lambda}{\lambda+\mu}\right].$$

6.6 Problems 6

6.1 (*Example 6.2.3*) Assuming $i = 1$ for the Yule process $\{N(t), t \geq 0\}$, we have the following Kolmogorov's forward equation:

$$P_1'(t) = -\lambda P_1(t),$$

$$P_k'(t) = -k\lambda P_k(t) + (k-1)\lambda P_{k-1}(t),$$

with $P_1(0) = 1$ and $P_k(0) = 0 \quad (k = 2, 3, \cdots)$. Let

$$M(t) = \sum_{i=1}^{\infty} iP_i(t)$$

be the mean of $N(t)$ at time t. Show that

$$M'(t) = \lambda M(t)$$

with $M(0) = 1$, and solve for $M(t)$.

6.2 (*Example 6.2.3*) Assuming $i = 1$ for the Yule process $\{N(t),\ t \geq 0\}$ (i.e., $N(0) = 1$ and $\lambda_k = k\lambda$), calculate the mean time and variance of $N(t) = n$. (Hint: Use the memoryless property of the exponential distribution).

6.3 (*Example 6.3.2*) Calculate the mean time and variance of $X(t) = k \quad (k < n)$ starting $X(0) = n$ at time 0. (Hint: Use the memoryless property of the exponential distribution).

6.4 (*Example 6.4.1*) Let

$$M^{(2)}(t) = E[X(t)^2] = \sum_{j=0}^{\infty} j^2 P_{ij}(t)$$

be the second moment of $X(t)$. Verify that

$$\frac{dM^{(2)}(t)}{dt} = 2(\lambda - \mu)M^{(2)}(t) + (\lambda + \mu)M(t),$$

where $M(t) = E[X(t)]$ is the mean of $X(t)$ in Example 6.4.1. Solve for $M^{(2)}(t)$ by assuming that $\lambda > \mu$, and derive the variance of $X(t)$.

6.5 (*Continuation*) Let

$$P(t, s) = \sum_{j=0}^{\infty} P_{ij}(t)s^j \qquad (|\, s \,| < 1)$$

be the generating function of $P_{ij}(t)$.

(i) Verify that $P(t, s)$ satisfies the following partial differential equation:

$$\frac{\partial P(t,s)}{\partial t} = [\lambda s^2 - (\lambda+\mu)s + \mu]\frac{\partial P(t,s)}{\partial s}.$$

(ii) Show that the generating function

$$P(t,s) = \frac{\mu(1-e^{(\lambda-\mu)t}) - (\lambda - \mu e^{(\lambda-\mu)t})s}{\mu - \lambda e^{(\lambda-\mu)t} - \lambda(1-e^{(\lambda-\mu)t})s}$$

satisfies the partial differential equation above with $X(0) = 1$, i.e., $P_{11}(0) = 1$, $P_{1j}(0) = 0$ $(j = 0, 2, 3, \cdots)$.

(iii) Expanding $P(t, s)$ in (ii) with respect to s, show that

$$P_{10}(t) = \alpha(t),$$

$$P_{1j}(t) = [1-\alpha(t)]\,[1-\beta(t)]\,[\beta(t)]^{j-1} \quad (j = 1, 2, \cdots),$$

and specify $\alpha(t)$ and $\beta(t)$.

(iv) Derive the probability that all members are dead as $t \to \infty$ by noting that state 0 is absorbing.

6.6 Consider a two-state birth and death process whose infinitesimal generator is

$$\mathbf{A} = \begin{matrix} 0 \\ 1 \end{matrix}\begin{bmatrix} -\lambda & \lambda \\ \mu & -\mu \end{bmatrix}.$$

Assume that the initial distribution vector is given by

$$\mathbf{p}(0) = \begin{bmatrix} \dfrac{\mu}{\lambda+\mu} & \dfrac{\lambda}{\lambda+\mu} \end{bmatrix}$$

which is the stationary distribution vector. Show that

$$\mathbf{p}(0)\mathbf{P}(t) = \mathbf{p}(0)e^{\mathbf{A}t} = \mathbf{p}(0).$$

That is, if the process starts with the stationary distribution $\mathbf{p}(0)$, then the process is stationary at any t.

6.7 Verify that there exists at least an eigenvalue $\omega = 0$ for a finite-state Markov chain with infinitesimal generator $\mathbf{A}$. In particular, the right eigenvector $\mathbf{p} = [p_i]$ for $\omega = 0$ corresponds to the stationary distribution, i.e.,

$$\mathbf{pA} = \mathbf{0} \text{ and } \sum_i p_i = 1.$$

6.8 Consider an item whose failure takes place either by type 1 failure with failure rate λ_1 or by type 2 failure with failure rate λ_2, whichever comes first. Once type i failure occurs, the repair is performed by the repair rate μ_i for type i failure ($i = 1, 2$). Upon repair the item recovers its function. Formulate a finite-state continuous-time Markov chain.

6.9 (*Linear Growth with Immigration*) Consider a birth and death process $\{X(t), t \geq 0\}$ with parameters $\{\lambda_k, \mu_{k+1}, k = 0, 1, 2, \cdots\}$, where

$$\lambda_k = k\lambda + a \quad \text{and} \quad \mu_{k+1} = (k+1)\mu$$

with $\lambda > 0$, $\mu > 0$ and $a > 0$. Such a birth and death process occurs in biological reproduction and population growth models where a is the immigration factor. Let

$$M(t) = E[X(t)] = \sum_{j=1}^{\infty} jP_{ij}(t)$$

be the mean of $X(t)$.

(i) Derive the differential equation for $M(t)$.

(ii) Solve the differential equation for $M(t)$ by classifying $\lambda = \mu$ or $\lambda \neq \mu$.

(iii) Discuss the limiting behavior of $M(t)$ as $t \to \infty$ by classifying $\lambda \geq \mu$ or $\lambda < \mu$.

6.10 (*Continuation*) Calculate the stationary distribution for the linear growth process with immigration by assuming that $\lambda < \mu$.

6.11 Consider a birth and death process $\{X(t), t \geq 0\}$ with parameters $\{\lambda_k, \mu_{k+1}, k = 0, 1, 2, \cdots\}$. Assume that

$$\lambda_k = \lambda, \; \mu_{k+1} = (k+1)\mu \quad (k = 0, 1, 2, \cdots).$$

Let

$$P(t, s) = \sum_{j=0}^{\infty} P_{0j}(t)s^j \quad (|\, s \,| < 1)$$

be the generating function of $P_{0j}(t)$.

(i) Derive the partial differential equation for $P(t, s)$.

(ii) Show that

$$P(t, s) = \exp\left[\frac{\lambda}{\mu}(1 - e^{-\mu t})(s - 1)\right]$$

is the solution for $P(t, s)$.

(iii) Expand $P(t,\ s)$ as a power series of j, and derive $P_{0j}(t)$.

(iv) Discuss the limiting probability of $P_{0j}(t)$ as $t \to \infty$.

6.12 (M/M/1 *Queue with Impatient Customers*) Consider an M/M/1 queue, where potential customers arrive at Poisson rate $\lambda_k = \lambda/(k+1)$ if there are k customers (including a customer in service) in a system, they are served exponentially at rate μ by a single channel, and the queue size is infinity. That is, the arrival rate λ_k is decreasing as k increases, since customers are impatient. Derive the limiting probabilities p_j $(j = 0, 1, 2, \cdots)$ and verify that they are the same for an M/M/∞ queue (see Example 6.4.3 and Section 9.3.3).

6.13 There are two machines, each of which fails exponentially at rate λ and is repaired immediately exponentially at rate μ. We assume that the two machines are operative at $t = 0$.

(i) Formulate a continuous-time Markov chain $\{X(t),\ t \geq 0\}$ by assuming that $X(0) = i$ $(i = 0, 1, 2)$ denotes that i machines are in repair at $t = 0$.

(ii) Derive the transition probabilities $P_{0j}(t)$ $(j = 0, 1, 2)$.

(iii) Derive the limiting transition probabilities $p_j = \lim_{t\to\infty} P_{0j}(t)$ $(j = 0, 1, 2)$.

6.14 (*Continuation*) Consider Problem 6.13 but now with only a single repair facility. That is, if two machines fail, one has to wait to be repaired.

(i) Formulate a continuous-time Markov chain.

(ii) Derive the limiting transition probabilities $p_j = \lim_{t\to\infty} P_{0j}(t)$ $(j = 0, 1, 2)$.

Chapter 7

Markov Renewal Processes

7.1 Introduction

We have discussed Markov chains in the preceding two chapters. In Chapter 5 we discussed discrete-time Markov chains in which the process can move from one state to another (including to itself) in discrete time. In Chapter 6 we discussed continuous-time Markov chains in which the process can move from one state to another, where each interarrival time is distributed exponentially. Note that only the exponential distribution has the memoryless property which plays an important role in analyzing the process.

Recall that we have discussed renewal processes in Chapter 4. We are concerned with the number of renewals (or events) in the renewal process, a process in which the interarrival time distributions are independent and identically arbitrary.

It is natural to combine a Markov chain and a renewal process. Such a process is called a Markov renewal process or a semi-Markov process. In this chapter we will discuss the Markov renewal process or semi-Markov process in general. In the following chapter, we will then develop the availability theory for system reliability models, which is a typical application of the Markov renewal process.

7.2 Markov Renewal Processes

Combining a Markov chain and a renewal process implies a *Markov renewal process* or a *semi-Markov process*. A Markov renewal process is concerned with generalized renewal random variables and a semi-Markov process is concerned with the random variables that the process is in a state at any time. However, the Markov renewal process and the semi-Markov process are equivalent from the viewpoint of probability theory. In the following we shall use the Markov renewal process.

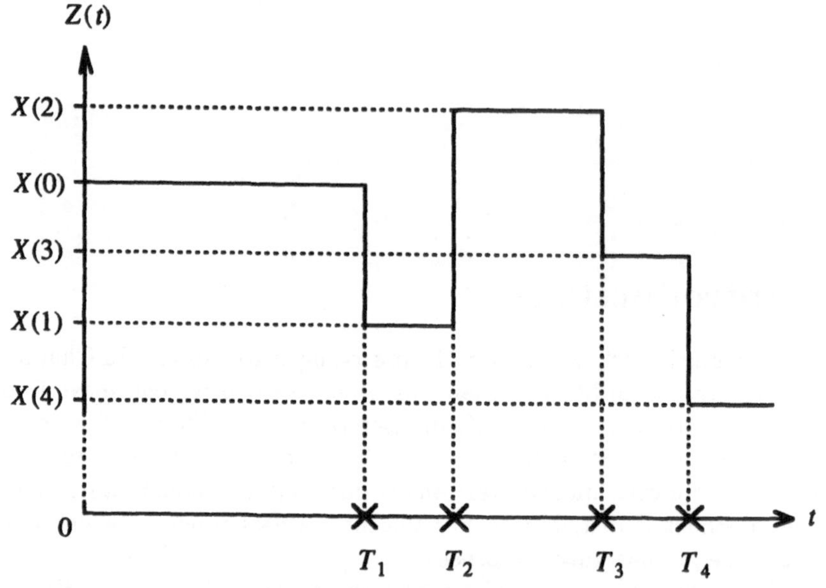

Fig. 7.2.1 A sample function of a stochastic process (**X**, **T**).

Consider a stochastic process (**X**, **T**), a sample function of which is shown in Fig. 7.2.1. Let the state space be $i = 0, 1, 2, \cdots, m$, where m is finite unless otherwise specified. We consider two vectorial random variables **X** and **T** in a Markov renewal process or a semi-Markov process. The vectorial random variable **X** is $\{X(n) : n = 0, 1, 2, \cdots\}$, where $X(n) = i$ denotes that the process is in state i at discrete-time n. That is, we are concerned with the discrete-time n which denotes the nth transition of the process ignoring the duration of the successive sojourn time (see Fig. 7.2.1). The vectorial random variable **T** is $\{T_n : n = 0, 1, 2, \cdots\}$, where T_n denotes the random variable of the nth arrival time at which the process just moves from one state to another (possibly to itself), where $T_0 = 0$ (see Fig. 7.2.1).

Let $\{Z(t), t \geq 0\}$ be a stochastic process, where $Z(t) = i$ denotes that the process is in state i at time t. Let $\{\mathbf{N}(t), t \geq 0\}$ be a stochastic process, where the vector $\mathbf{N}(t)$ is given by

$$\mathbf{N}(t) = [N_0(t)\ N_1(t) \cdots\ N_m(t)], \tag{7.2.1}$$

and $N_i(t) = k$ denotes that the random number of visits to state i is k in the time interval $(0, t]$. The random variable $Z(t)$ can specify the state at time t, and the vectorial random variable $\mathbf{N}(t)$ can specify the generalized renewal random variables for all the states $i = 0, 1, 2, \cdots, m$. We are now in a position to define a Markov renewal process and a semi-Markov process.

Definition 7.2.1 A stochastic process $\{\mathbf{N}(t), t \geq 0\}$ is called a *Markov renewal process* if

$$\begin{aligned} &P\{X(n+1) = j, T_{n+1} - T_n \leq t \mid \\ &X(0) = i_0, \cdots, X(n) = i; T_0 = 0, T_1 = t_1, \cdots, T_n = t_n\} \\ &= P\{X(n+1) = j, T_{n+1} - T_n \leq t \mid X(n) = i, T_n = t_n\} \\ &= Q_{ij}(t) \end{aligned} \tag{7.2.2}$$

is satisfied for all $n = 0, 1, 2, \cdots; i, j = 0, 1, 2, \cdots, m$ and $t \in [0, \infty)$.

Definition 7.2.2 A stochastic process $\{Z(t), t \geq 0\}$ is called a *semi-Markov process* if Eq.(7.2.2) is satisfied for all $n = 0, 1, 2, \cdots; i, j = 0, 1, 2, \cdots, m$ and $t \in [0, \infty)$.

Roughly speaking, both the Markov renewal process and the semi-Markov process are the the same from the viewpoint of stochastic processes. However, the former is concerned with the vectorial random variable $\mathbf{N}(t)$ and the latter with the random variable $Z(t)$. We shall cite several examples where we are interested in both random variables $Z(t)$ and $\mathbf{N}(t)$ in the analysis of a model. We shall use the Markov renewal process for convenience.

Of course, we assume that the process is time-homogeneous, i.e.,

$$Q_{ij}(t) = P\{X(n+1) = j,\ T_{n+1} - T_n \leq t \mid X(n) = i,\ T_n = t_n\} \tag{7.2.3}$$

is independent of $T_n = t_n$. Then $Q_{ij}(t)$ is called a *mass function* or a *one-step transition probability* for the Markov renewal process. The matrix $\mathbf{Q}(t)$ composed of $Q_{ij}(t)$, i.e.,

$$\mathbf{Q}(t) = [Q_{ij}(t)] \tag{7.2.4}$$

is called a *semi-Markov kernel.* The one-step transition probability $Q_{ij}(t)$ is the probability that, after making a transition into state i, the process next makes a transition into state j (possibly $i = j$) in an amount of time less than or equal to t. The one-step transition probabilities satisfy the following:

$$Q_{ij}(t) \geq 0, \ \sum_{j=0}^{m} Q_{ij}(\infty) = 1 \qquad (i, j = 0, 1, 2, \cdots, m). \tag{7.2.5}$$

Let

$$p_{ij} = \lim_{t \to \infty} Q_{ij}(t) = P\{X(n+1) = j \mid X(n) = i\} \tag{7.2.6}$$

denote the *eventual transition probability* that the process can move from state i to state j, neglecting the sojourn time in state i. That is, p_{ij} is the eventual transition probability, neglecting the sojourn time, which governs the behavior of the random variable $\{X(n),\ n = 0, 1, 2, \cdots\}$. The eventual transition probabilities satisfy the following:

$$p_{ij} \geq 0 \qquad (i, j = 0, 1, 2, \cdots, m). \tag{7.2.7}$$

$$\sum_{j=0}^{m} p_{ij} = 1 \qquad (i = 0, 1, 2, \cdots, m). \tag{7.2.8}$$

A discrete-time Markov chain $\{X(n),\ n = 0, 1, 2, \cdots\}$ having the eventual transition probabilities p_{ij} $(i, j = 0, 1, 2, \cdots, m)$ is called an *embedded Markov chain*. If we assume only well-behaved distributions, the behavior of the Markov renewal process is governed by the embedded Markov chain by ignoring all the sojourn times among the state transitions.

If $p_{ij} > 0$ for some i and j, then we can define

$$F_{ij}(t) = Q_{ij}(t)/p_{ij}, \tag{7.2.9}$$

and if $p_{ij} = 0$ for some i and j, then we can define $Q_{ij}(t) = 0$ for all $t \geq 0$ and $F_{ij}(t) = 1(t)$ (a step function). The distribution $F_{ij}(t)$ is a distribution of the sojourn time that the process spends in state i given that the next visiting state is j.

Example 7.2.1 (*Renewal Process*) Consider a Markov renewal process $\{N_0(t), t \geq 0\}$ with a single state, say 0, where $p_{00} = 1$ and $F_{00}(t)$ is an arbitrary distribution. That is, such a Markov renewal process is a renewal process with the interarrival time distribution $F_{00}(t)$, which has been thoroughly discussed in Chapter 4.

Example 7.2.2 (*Discrete-Time Markov Chain*) Consider a semi-Markov process $\{Z(t), t \geq 0\}$, where we assume that $Q_{ij}(t) = p_{ij}1(t-1)$ for all $i, j = 0, 1, \cdots, m$, where $1(t-1)$ is a step function at $t = 1$. Then such a semi-Markov process $\{Z(t), t \geq 0\}$ is a discrete-time Markov chain $\{X(n), n = 0, 1, 2, \cdots\}$ with transition probabilities p_{ij} $(i, j = 0, 1, 2, \cdots, m)$.

Example 7.2.3 (*Continuous-Time Markov Chain*) Consider a semi-Markov process $\{Z(t), t \geq 0\}$, where we assume that

$$Q_{ij}(t) = p_{ij}(1 - e^{-\lambda_i t})$$

if $p_{ij} > 0$, and

$$Q_{ij}(t) = 0$$

if $p_{ij} = 0$ for all $i, j = 0, 1, 2, \cdots, m$; $i \neq j$ and $Q_{ii}(t) = 0$ $(i = 0, 1, 2, \cdots, m)$. Such a semi-Markov process is a continuous-time Markov chain with the infinitesimal generator $\mathbf{A}$ with its non-diagonal elements $p_{ij}\lambda_i$ $(i \neq j)$ and diagonal elements $-\lambda_i$ $(i,\ j = 0, 1, 2, \cdots, m)$.

Example 7.2.4 (*Two-State Birth and Death Process*) Consider a two-state birth and death process as in Example 6.4.4. As shown in Example 7.2.3, we can describe such a process by assuming

$$Q_{01}(t) = 1 - e^{-\lambda t}, \; Q_{10}(t) = 1 - e^{-\mu t}, \; Q_{00}(t) = Q_{11}(t) = 0.$$

As shown in the above examples, a Markov renewal process or a semi-Markov process includes a renewal process, a discrete-time Markov chain, and a continuous-time Markov chain as its special cases. That is, a Markov renewal process, which is a generalization of a renewal process, a discrete-time Markov chain, and a continuous-time Markov chain, can permit the arbitrary sojourn time distributions $F_{ij}(t)$ for all i and j, and satisfies the Markov property in Eq.(7.2.2) in a sense. The Markov property in Eq.(7.2.2) shows us that the process is governed by the elapsed time from the latest time instant at which a transition (i.e., an event or occurrence) takes place. Such a time instant at which a transition takes place is called a *regeneration point* since we cannot take account of the elapsed time at the regeneration point (i.e., the Markov property is satisfied only at the regeneration points in general). Recall that a continuous-time stochastic process such as a Poisson process or a continuous-time Markov chain satisfies the Markov property at all $t \geq 0$. However, a Markov renewal process satisfies the Markov property in Eq.(7.2.2) only at the regeneration points. Proceeding to the successive regeneration points (i.e., $T_n, n = 1, 2, \cdots$, in Fig. 7.2.1), we can analyze a Markov renewal process by applying and expanding the existing renewal and Markov chain theories, since the behavior of the process can be determined by the latest regeneration point which can specify the sojourn time and the next state visited.

All the states of a Markov renewal process can be classified by using the corresponding embedded Markov chain. However, the periodicity of the embedded Markov chain has nothing to do with that of the underlying Markov renewal process. For instance, if the embedded Markov chain is irreducible and periodic, and every $Q_{ij}(t)$ is continuous in time t, we can consider that the underlying Markov renewal process is aperiodic since $Q_{ij}(t)$ is continuous (or non-lattice) in time t. We find an exception to the Markov renewal process. If the embedded Markov chain is irreducible and every $Q_{ij}(t)$ $(i, j = 0, 1, \cdots, m)$ is lattice with common period δ, then such a Markov renewal process is called a discrete-time Markov renewal process and has some applications in applied stochastic models. However, we do not intend to discuss such a Markov renewal process. We assume that a Markov renewal process is aperiodic in general throughout this chapter.

If the sojourn time $F_{ij}(t)$ is distributed exponentially, then the process satisfies the Markov property in Eq.(7.2.2). In this case, the process that stays in state i is independent of the elapsed time measured from the time instant at which a transition to state i takes place because of the memoryless property of the exponential distribution. However, such a time instant is no longer a regeneration point. A regeneration point is a time instant at which a transition take place and that satisfies the Markov property in Eq.(7.2.2). Since we are concerned with the generalized random variable $\mathbf{N}(t)$ which can count the numbers of events, occurrences, or renewals in the process for the time interval $(0, t]$, we have to specify the regeneration point at which not only the Markov property in Eq.(7.2.2) is satisfied, but also an event, occurrence, or renewal takes place (refer to the Poisson process in Chapter 3).

Recall that $Q_{ij}(t) = p_{ij}F_{ij}(t)$ is a one-step transition probability from state i to state j for $i,\ j = 0, 1, 2, \cdots, m$. Assume that the first and second moments of the sojourn time distribution $F_{ij}(t)$ exist and are given by

$$\nu_{ij} = \int_0^\infty t\,dF_{ij}(t), \quad \nu_{ij}^{(2)} = \int_0^\infty t^2\,dF_{ij}(t) \qquad (i, j = 0, 1, 2, \cdots, m), \tag{7.2.10}$$

respectively. Define

$$H_i(t) = \sum_{j=0}^{m} Q_{ij}(t) \qquad (i = 0, 1, 2, \cdots, m), \tag{7.2.11}$$

which is called the *uncondional distribution* in state i since $H_i(t)$ is the distribution not specifying the next state visited. We also define the first and second moments of $H_i(t)$:

$$\xi_i = \int_0^\infty t\,dH_i(t) = \sum_{j=0}^{m} p_{ij}\nu_{ij}, \quad \xi_i^{(2)} = \int_0^\infty t^2\,dH_i(t) = \sum_{j=0}^{m} p_{ij}\nu_{ij}^{(2)}$$

$$(i = 0, 1, 2, \cdots, m), \tag{7.2.12}$$

respectively. In particular, ξ_i is called the *unconditional mean* in state i.

Define the *Markov renewal function*

$$M_{ij}(t) = E[N_j(t) \mid Z(0) = i] \qquad (i, j = 0, 1, 2, \cdots, m), \tag{7.2.13}$$

which is the generalized renewal function and is the expected number of visits to state j in an interval $(0, t]$ given that the process started in state i at time 0. Combining renewal and Markov chain theories implies

$$\begin{aligned} M_{ij}(t) &= \sum_{\substack{k=0 \\ k \neq j}}^{m} \int_0^t M_{kj}(t-x)\,dQ_{ik}(x) + \int_0^t [1 + M_{jj}(t-x)]\,dQ_{ij}(x) \\ &= Q_{ij}(t) + \sum_{k=0}^{m} \int_0^t M_{kj}(t-x)\,dQ_{ik}(x) \\ &= Q_{ij}(t) + \sum_{k=0}^{m} Q_{ik} * M_{kj}(t), \end{aligned} \tag{7.2.14}$$

where the notation $*$ denotes the Stieltjes convolution in Eq.(2.5.27). Introducing the matrices $\mathbf{M(t)} = [M_{ij}(t)]$ and $\mathbf{Q}(t) = [Q_{ij}(t)]$ in Eq.(7.2.4), we can rewrite Eq.(7.2.14) in the matrix form:

$$\mathbf{M}(t) = \mathbf{Q}(t) + \mathbf{Q} * \mathbf{M}(t), \tag{7.2.15}$$

where the notation $*$ in the matrix denotes matrix multiplication except that the multiplication of each element is replaced by the Stieltjes convolution of each element. Eq.(7.2.15) is a *renewal equation* in matrix form and is a direct extension of Eq.(4.2.11).

Equation (7.2.15) can be rewritten as

$$[\mathbf{I} - \mathbf{Q}] * \mathbf{M}(t) = \mathbf{Q}(t), \tag{7.2.16}$$

where $\mathbf{I}$ is an identity matrix with the diagonal elements $1(t)$ (step functions). Define the inverse of the matrix $\mathbf{I} - \mathbf{Q}(t)$ as

$$[\mathbf{I} - \mathbf{Q}(t)]^{(-1)*} = \sum_{n=0}^{\infty} \mathbf{Q}^{n*}(t), \tag{7.2.17}$$

where $\mathbf{Q}^{n*}(t)$ is the nth power of the matrix product of $\mathbf{Q}(t)$ except that the multiplication of each element is replaced by the Stieltjes convolution of each element. Noting that the inverse $[\mathbf{I} - \mathbf{Q}(t)]^{(-1)*}$ exists for a finite t, we can solve Eq.(7.2.16) for $\mathbf{M}(t)$:

$$\begin{aligned} \mathbf{M}(t) &= [\mathbf{I} - \mathbf{Q}]^{(-1)*} * \mathbf{Q}(t) \\ &= \sum_{n=1}^{\infty} \mathbf{Q}^{n*}(t) \\ &= [\mathbf{I} - \mathbf{Q}(t)]^{(-1)*} - \mathbf{I}, \end{aligned} \tag{7.2.18}$$

which corresponds to Eq.(4.2.11) in matrix form.

Let $P_{ij}(t)$ denote the transition probability that the process is in state j at time t given that it was in state i at time 0:

$$P_{ij}(t) = P\{Z(t) = j \mid Z(0) = i\}. \tag{7.2.19}$$

Let $G_{ij}(t)$ denote the first passage time distribution that the process first arrives at state j up to time t given that it was in state i at time 0:

$$G_{ij}(t) = P\{N_j(t) > 0 \mid Z(0) = i\}. \tag{7.2.20}$$

The transition probability $P_{ij}(t)$ can be expressed in terms of $Q_{ij}(t)$ and $H_i(t)$:

$$P_{ij}(t) = [1 - H_i(t)]\delta_{ij} + \sum_{k=0}^{m} Q_{ik} * P_{kj}(t). \tag{7.2.21}$$

That is, if $i \neq j$, the process can move to state k and then move to state j up to time t. If $i = j$, there is another possibility that the process cannot move from state i up to time t, which can be expressed in the first term of the right-hand side of Eq.(7.2.21), where δ_{ij} is a *Kronecker's delta*, i.e., $\delta_{ij} = 1$ if $i = j$ and $\delta_{ij} = 0$ if $i \neq j$.

The transition probability $P_{ij}(t)$ can also be expressed in terms of $H_i(t)$ and $G_{ij}(t)$:

$$P_{ij}(t) = [1 - H_i(t)]\delta_{ij} + G_{ij} * P_{jj}(t), \tag{7.2.22}$$

which can be similarly interpreted. That is, if $i \neq j$, the process can first move to state j and then follow the transition probability $P_{jj}(t)$. If $i = j$, there is the other possibility that the process can stay in state i up to time t.

The first passage time distribution $G_{ij}(t)$ can be expressed in terms of $Q_{ij}(t)$:

$$G_{ij}(t) = Q_{ij}(t) + \sum_{\substack{k=0 \\ k \neq j}}^{m} Q_{ik} * G_{kj}(t), \tag{7.2.23}$$

which can be similarly interpreted.

Let us introduce the Laplace-Stieltjes transforms of $Q_{ij}(t)$ and $M_{ij}(t)$:

$$Q_{ij}^*(s) = \int_0^\infty e^{-st}\, dQ_{ij}(t) \qquad (i, j = 0, 1, 2, \cdots, m), \tag{7.2.24}$$

$$M_{ij}^*(s) = \int_0^\infty e^{-st}\, dM_{ij}(t) \qquad (i, j = 0, 1, 2, \cdots, m). \tag{7.2.25}$$

Introducing the matrices $\mathbf{Q}^*(s) = [Q_{ij}^*(s)]$ and $\mathbf{M}^*(s) = [M_{ij}^*(s)]$, we can express Eq.(7.2.15) in matrix form by using the Laplace-Stieltjes transforms:

$$\mathbf{M}^*(s) = \mathbf{Q}^*(s) + \mathbf{Q}^*(s)\mathbf{M}^*(s). \tag{7.2.26}$$

Introducing an identity matrix $\mathbf{I}$ and noting that the inverse $[\mathbf{I} - \mathbf{Q}^*(s)]^{-1}$ exists for $\Re(s) > 0$, we have

$$\begin{aligned} \mathbf{M}^*(s) &= [\mathbf{I} - \mathbf{Q}^*(s)]^{-1}\mathbf{Q}^*(s) \\ &= [\mathbf{I} - \mathbf{Q}^*(s)]^{-1} - \mathbf{I}, \end{aligned} \tag{7.2.27}$$

which corresponds to Eq.(4.2.15) in matrix form. Once $\mathbf{Q}(t)$ is given, we can obtain the matrix Laplace-Stieltjes transform $\mathbf{Q}^*(s)$ and its inverse $[\mathbf{I} - \mathbf{Q}^*(s)]^{-1}$. That is, $\mathbf{M}^*(s)$ can be given in Eq.(7.2.27).

Let us introduce the Laplace-Stieltjes transforms of $G_{ij}(t)$ and $P_{ij}(t)$:

$$G_{ij}^*(s) = \int_0^\infty e^{-st}\, dG_{ij}(t) \qquad (i, j = 0, 1, 2, \cdots, m), \tag{7.2.28}$$

$$P_{ij}^*(s) = \int_0^\infty e^{-st}\, dP_{ij}(t) \qquad (i, j = 0, 1, 2, \cdots, m). \tag{7.2.29}$$

Noting that $M_{ij}(t) = G_{ij} * [1 + M_{jj}](t)$, we have the Laplace-Stieltjes transforms

$$M_{ij}(s) = G_{ij}^*(s) + G_{ij}^*(s)M_{jj}^*(s), \tag{7.2.30}$$

which implies

$$G_{ij}^*(s) = \frac{M_{ij}^*(s)}{1 + M_{jj}^*(s)} \qquad (i, j = 0, 1, 2, \cdots, m). \tag{7.2.31}$$

From Eq.(7.2.22), we have

$$P_{jj}^*(s) = \frac{1 - H_j^*(s)}{1 - G_{jj}^*(s)} \qquad (j = 0, 1, 2, \cdots, m), \tag{7.2.32}$$

$$P_{ij}^*(s) = G_{ij}^*(s)P_{jj}^*(s) \quad (i, j = 0, 1, 2, \cdots, m; i \neq j), \tag{7.2.33}$$

where $H_i^*(s) = \sum_{j=0}^{m} Q_{ij}^*(s)$. We can analytically obtain the Laplace-Stieltjes transforms $M_{ij}^*(s)$ in Eq.(7.2.27), $G_{ij}^*(s)$ in Eq.(7.2.31), and $P_{ij}^*(s)$ in Eqs.(7.2.32) and (7.2.33), respectively. In principle, we can invert the corresponding Laplace-Stieltjes transforms $M_{ij}^*(s)$, $G_{ij}^*(s)$, and $P_{ij}^*(s)$, which imply the analytical forms of $M_{ij}(t)$, $G_{ij}(t)$, and $P_{ij}(t)$. However, it is very difficult or impossible to do so except in the simplest cases.

Example 7.2.5 (*Example 7.2.4*) Consider a two-state birth and death process. Then

$$\mathbf{Q}^*(s) = \begin{bmatrix} 0 & \dfrac{\lambda}{s+\lambda} \\ \dfrac{\mu}{s+\mu} & 0 \end{bmatrix},$$

and

$$\mathbf{M}^*(s) = [\mathbf{I} - \mathbf{Q}^*(s)]^{-1} - \mathbf{I} = \frac{1}{s(s+\lambda+\mu)} \begin{bmatrix} \lambda\mu & \lambda(s+\mu) \\ \mu(s+\lambda) & \lambda\mu \end{bmatrix}$$

which implies by inversion

$$M_{00}(t) = M_{11}(t) = \frac{\lambda\mu t}{\lambda+\mu} - \frac{\lambda\mu}{(\lambda+\mu)^2}\left[1 - e^{-(\lambda+\mu)t}\right],$$

$$M_{01}(t) = \frac{\lambda\mu t}{\lambda+\mu} + \frac{\lambda^2}{(\lambda+\mu)^2}\left[1 - e^{-(\lambda+\mu)t}\right],$$

$$M_{10}(t) = \frac{\lambda\mu t}{\lambda+\mu} + \frac{\mu^2}{(\lambda+\mu)^2}\left[1 - e^{-(\lambda+\mu)t}\right].$$

From Eq.(7.2.31), we have

$$G_{01}^*(s) = \frac{\lambda}{s+\lambda}, \; G_{10}^*(s) = \frac{\mu}{s+\mu}, \; G_{jj}^*(s) = \frac{\lambda\mu}{(s+\lambda)(s+\mu)} \qquad (j = 0, 1),$$

which imply

$$G_{01}(t) = 1 - e^{-\lambda t}, \; G_{10}(t) = 1 - e^{-\mu t},$$

$$G_{jj}(t) = \begin{cases} \dfrac{\mu}{\mu-\lambda}(1 - e^{-\lambda t}) + \dfrac{\lambda}{\lambda-\mu}(1 - e^{-\mu t}) & (\lambda \neq \mu) \\ 1 - (1 + \lambda t)e^{-\lambda t} & (\lambda = \mu). \end{cases}$$

From Eqs.(7.2.32) and (7.2.33) we also have

$$P_{00}^*(s) = \frac{s+\mu}{s+\lambda+\mu}, \quad P_{11}^*(s) = \frac{s+\lambda}{s+\lambda+\mu},$$

$$P_{01}^*(s) = \frac{\lambda}{s+\lambda+\mu}, \quad P_{10}^*(s) = \frac{\mu}{s+\lambda+\mu},$$

which, by inversion, imply $P_{ij}(t) \quad (i,j=0,1)$ in Example 6.4.5.

7.3 Stationary Probabilities

In the preceding section we have developed the Markov renewal process and obtained the Laplace-Stieltjes transforms $M_{ij}^*(s)$, $G_{ij}^*(s)$, and $P_{ij}^*(s) \quad (i,j=0,1,2,\cdots,m)$, which imply the corresponding probabilistic quantities $M_{ij}(t)$, $G_{ij}(t)$, and $P_{ij}(t)$ by inversion. However, it is very difficult or impossible to obtain such probabilistic quantities analytically. In this section we shall discuss the stationary (or limiting) probabilities for Markov renewal processes.

Let μ_{ij} and $\mu_{ij}^{(2)}$ denote the first and second moments about the origin of $G_{ij}(t)$, respectively. From the probabilistic arguments, we have

$$\begin{aligned} \mu_{ij} &= \sum_{k\neq j} p_{ik}(\nu_{ik} + \mu_{kj}) + p_{ij}\nu_{ij} \\ &= \xi_i + \sum_{k\neq j} p_{ik}\mu_{kj}, \end{aligned} \tag{7.3.1}$$

and

$$\mu_{ij}^{(2)} = \xi_i^{(2)} + \sum_{k\neq j} p_{ik}\left[\mu_{kj}^{(2)} + 2\nu_{ik}\mu_{kj}\right], \tag{7.3.2}$$

where ξ_i and $\xi_i^{(2)}$ are the first and second moments of $H_i(t)$, which are given in Eq.(7.2.12), and ν_{ij} is the mean of $F_{ij}(t)$, which is given in Eq.(7.2.10).

If the embedded Markov chain is irreducible, there exists a unique and positive probability vector $\pi = [\pi_j]$, where π is the stationary distribution (vector) and satisfies

$$\pi\mathbf{P} = \pi, \quad \sum_{i=0}^{m} \pi_i = 1. \tag{7.3.3}$$

Note that $\mathbf{P} = [p_{ij}]$ is the transition probability matrix of the *embedded* Markov chain. Multiplying both sides by π_i and summing over i in Eqs.(7.3.1) and (7.3.2), we have

$$\sum_{i=0}^{m} \pi_i\mu_{ij} = \sum_{i=0}^{m}\sum_{k\neq j} \pi_i p_{ik}\mu_{kj} + \sum_{i=0}^{m} \pi_i\xi_i$$

$$= \sum_{k \neq j} \pi_k \mu_{kj} + \sum_{i=0}^{m} \pi_i \xi_i, \tag{7.3.4}$$

which implies

$$\mu_{jj} = \frac{1}{\pi_j} \sum_{i=0}^{m} \pi_i \xi_i \qquad (j = 0, 1, 2, \cdots, m), \tag{7.3.5}$$

and

$$\sum_{i=0}^{m} \pi_i \mu_{ij}^{(2)} = \sum_{i=0}^{m} \pi_i \xi_i^{(2)} + \sum_{i=0}^{m} \sum_{k \neq j} \pi_i p_{ik} \left[\mu_{kj}^{(2)} + 2\nu_{ik} \mu_{kj} \right]$$

$$= \sum_{i=0}^{m} \pi_i \xi_i^{(2)} + \sum_{k \neq j} \pi_k \mu_{kj}^{(2)} + 2 \sum_{k \neq j} \sum_{i=0}^{m} \pi_i p_{ik} \nu_{ik} \mu_{kj}, \tag{7.3.6}$$

which implies

$$\mu_{jj}^{(2)} = \frac{1}{\pi_j} \left[\sum_{i=0}^{m} \pi_i \xi_i^{(2)} + 2 \sum_{k \neq j} \sum_{i=0}^{m} \pi_i p_{ik} \nu_{ik} \mu_{kj} \right]. \tag{7.3.7}$$

Example 7.3.1 (*Discrete-Time Markov Chain*) For a finite-state discrete-time Markov chain, if the process is irreducible and aperiodic, we have $\xi_i = 1$, $\xi_i^{(2)} = 1$, $\nu_{ik} = 1$, $\pi_i > 0$ $(i, k = 0, 1, 2, \cdots, m)$, and $\sum_{i=0}^{m} \pi_i = 1$, i.e.,

$$\mu_{jj} = \frac{1}{\pi_j} \qquad (j = 0, 1, 2, \cdots, m),$$

$$\mu_{jj}^{(2)} = \frac{1}{\pi_j} \left[1 + 2 \sum_{k \neq j} \pi_k \mu_{kj} \right] \qquad (j = 0, 1, 2, \cdots, m).$$

The probability $G_{ij}(\infty)$ that the process is eventually in state j given that it was in state i at time 0 is given by

$$G_{ij}(\infty) = \begin{cases} 1 & (i, j \in C_k;\ k = 1, 2, \cdots, K) \\ 0 & (i \in C_k,\ j \in C_l;\ k \neq l) \\ f_{ij} & (i \in T,\ j \in C_k;\ k = 1, 2, \cdots, K), \end{cases} \tag{7.3.8}$$

where f_{ij} is given in Eq.(5.5.6) or (5.5.7) for a finite-state discrete-time Markov chain whose transition probability matrix is given by $\mathbf{P} = [p_{ij}]$, i.e., that of the embedded Markov chain, and whose state can be classified into some recurrent classes $C_1, C_2, \cdots, C_K$ and a set T of the remaining transient states (see Eq.(5.5.1)). These results are just the same as those of the discrete-time Markov chain theory developed in Section 5.5 since we are concerned with the transition probability and can neglect the sojourn time in each state.

We are now in a position to derive the stationary probabilities for a Markov renewal process. By first noting that the stationary probabilities π_j ($j = 0, 1, 2, \cdots$, m) of the embedded Markov chain can be given by the discrete-time Markov chain theory developed in Section 5.4, we can then derive the stationary probabilities for the Markov renewal process. If we assume that the Markov renewal process is irreducible and aperiodic, we can apply the Tauberian theorem (see Appendix A) to Eq.(7.2.32):

$$\lim_{t\to\infty} P_{jj}(t) = \lim_{s\to 0} P_{jj}^*(s) = \lim_{s\to 0} \frac{[1 - H_j^*(s)]/s}{[1 - G_{jj}^*(s)]/s} = \frac{\xi_j}{\mu_{jj}}. \tag{7.3.9}$$

In general, we have from Eq.(7.2.33):

$$\lim_{t\to\infty} P_{ij}(t) = \lim_{s\to 0} G_{ij}^*(s) P_{jj}^*(s) = \frac{G_{ij}(\infty)\xi_j}{\mu_{jj}}. \tag{7.3.10}$$

Summarizing the above results, we have the following:

Theorem 7.3.1 The stationary probabilities ρ_j $(j = 0, 1, 2, \cdots, m)$ for a finite-state Markov renewal process are given by

$$\rho_j = \begin{cases} \dfrac{\xi_j}{\mu_{jj}} & (j \in C_k;\ k = 1, 2, \cdots, K), \\ \dfrac{G_{ij}(\infty)\xi_j}{\mu_{jj}} & (i \in T,\ j \in C_k;\ k = 1, 2, \cdots, K), \\ 0 & (otherwise), \end{cases} \tag{7.3.11}$$

if $\xi_j < \infty$ for all $j = 0, 1, 2, \cdots, m$. In particular, if the process is irreducible and aperiodic, we have from Eq.(7.3.5):

$$\rho_j = \frac{\xi_j}{\mu_{jj}} = \frac{\pi_j \xi_j}{\sum_{k=0}^m \pi_k \xi_k} \qquad (j = 0, 1, 2, \cdots, m). \tag{7.3.12}$$

Let $T_{N(t)}$ be the arrival time of the latest transition before or at time t and $T_{N(t)+1}$ be the next arrival time after t. Then the random variable $\gamma(t) = T_{N(t)+1} - t$ and $\delta(t) = t - T_{N(t)}$ are of interest, where $\gamma(t)$ and $\delta(t)$ are the generalized *excess life* (*residual life*) and *current life* (*age*) random variables, respectively (see Fig. 3.4.3). If we assume that the underlying Markov renewal process is irreducible and aperiodic, and the mean recurrence times μ_{jj} are finite, we have the following limiting results:

$$\begin{aligned} &\lim_{t\to\infty} P\{\delta(t) \le x \mid Z(t) = i\} \\ &= \lim_{t\to\infty} P\{\gamma(t) \le x \mid Z(t) = i\} \\ &= \frac{1}{\xi_i}\int_0^x [1 - H_i(u)]\, du, \end{aligned} \tag{7.3.13}$$

which generalizes the well-known result in Eqs.(4.3.43) and (4.3.38) for a renewal process.

Let us define the following probability:

$$ {}_iR_{jk}(x;t) = P\left\{Z(t)=j,\ X(N(t)+1)=k,\ T_{N(t)+1}-t \le x \mid Z(0)=i\right\} $$
$$ (i,j,k = 0,1,2,\cdots,m). \tag{7.3.14} $$

We are concerned with the limiting probability in Eq.(7.3.14) as $t \to \infty$. We provide the following theorem without proof:

Theorem 7.3.2 If state j is recurrent and aperiodic, and $\nu_{jk} < \infty$, then

$$ \lim_{t\to\infty} {}_iR_{jk}(x;t) = \frac{G_{ij}(\infty)p_{jk}}{\mu_{jj}} \int_0^x [1-F_{jk}(y)]\,dy. \tag{7.3.15} $$

The *stationary Markov renewal process* can be constructed from the stationary probabilities $\rho_j (i = 0,1,2,\cdots,m)$ in Eq.(7.3.11) and $\lim_{t\to\infty} {}_iR_{jk}(x;t)$ in Eq.(7.3.15) if we assume that the process is irreducible and aperiodic. That is, if we assume that the process is irreducible and aperiodic, we can construct the following process:

$$ P\{Z(0)=j\} = \rho_j \quad (j=0,1,2,\cdots,m), \tag{7.3.16} $$

$$ P\{X(1)=j,\ T_1-T_0 \le x \mid X(0)=i\} $$
$$ = \frac{p_{ij}}{\xi_i}\int_0^x [1-F_{ij}(y)]dy, \tag{7.3.17} $$

$$ P\{X(n+1)=j,\ T_{n+1}-T_n \le x \mid X(n)=i\} = Q_{ij}(x) $$
$$ (n=1,2,\cdots). \tag{7.3.18} $$

We can construct the stationary Markov renewal process which is a direct expansion of the stationary renewal process discussed in Section 4.4 .

Let us discuss the asymptotic behavior of $M_{ij}(t)$. From Eq.(7.2.31), we have

$$ \begin{aligned} M^*_{ij}(s) &= G^*_{ij}(s)[1+M^*_{jj}(s)] \\ &= G^*_{ij}(s)\left[1+\frac{G^*_{jj}(s)}{1-G^*_{jj}(s)}\right]. \end{aligned} \tag{7.3.19} $$

Noting that

$$ G^*_{ij}(s) = 1 - s\mu_{ij} + \frac{s^2}{2!}\mu^{(2)}_{ij} + o(s^2), \tag{7.3.20} $$

and substituting Eq.(7.3.20) into Eq.(7.3.19), we have

$$ M^*_{ij}(s) = \frac{1}{s\mu_{jj}} + \frac{\mu^{(2)}_{jj}}{2\mu^2_{jj}} - \frac{\mu_{ij}}{\mu_{jj}} + o(1). \tag{7.3.21} $$

If we further assume that states i and j belong to the same recurrent class, and apply the Tauberian theorem to Eq.(7.3.21), we have

$$ M_{ij}(t) - \frac{t}{\mu_{jj}} \longrightarrow \frac{\mu^{(2)}_{jj}}{2\mu^2_{jj}} - \frac{\mu_{ij}}{\mu_{jj}} \tag{7.3.22} $$

as $t \to \infty$, which can be obtained by analogy with a delayed renewal process since the first interarrival time is distributed with $G_{ij}(t)$.

7.4 Alternating Renewal Processes

We have developed Markov renewal processes in general in the preceding two sections. In this section we discuss an *alternating renewal process* as one of the simplest examples of Markov renewal processes. An alternating renewal process is a Markov renewal process with two states and has fruitful results for the probabilistic quantities developed in the preceding two sections in terms of the Laplace-Stieltjes transforms. Of course, an alternating renewal process can be analyzed by expanding the renewal-theoretic arguments. However, in this section we analyze the alternating renewal process as an example of Markov renewal processes. The results obtained in this section are of direct use in developing availability theory for a one-unit system in the following chapter (see Section 8.3).

Consider a Markov renewal process whose state space is composed of two states, say, $i = 0, 1$, where, for the convenience of the following chapter, state 0 is a failed state and state 1 is an operating state. The process can move from one state to another according to arbitrary distributions. Let $\{\mathbf{N}(t), t \geq 0\}$ be a Markov renewal process (i.e., an alternating renewal process) whose semi-Markov kernel is given by

$$\mathbf{Q}(t) = \begin{matrix} 0 \\ 1 \end{matrix} \begin{bmatrix} 0 & G(t) \\ F(t) & 0 \end{bmatrix}, \tag{7.4.1}$$

where $Q_{01}(t) = G(t)$ and $Q_{10}(t) = F(t)$ are arbitrary distributions of failed and operating states, respectively. It is clear that the embedded Markov chain is given by

$$\mathbf{P} = \begin{matrix} 0 \\ 1 \end{matrix} \begin{bmatrix} 0 & 1 \\ 1 & 0 \end{bmatrix}. \tag{7.4.2}$$

Let X and Y denote the interarrival times for each state. Then, the distribution of $X + Y$ is given by $G * F(t)$ and plays a central role in the analysis of the process. By using the n-fold convolution

$$(G * F)^{n*}(t) = \overbrace{G * F * G * F * \cdots * G * F(t)}^{2n}, \tag{7.4.3}$$

we have the Markov renewal functions in the matrix form:

$$\mathbf{M}(t) = \begin{bmatrix} \sum_{n=1}^{\infty}(G * F)^{n*}(t) & \sum_{n=0}^{\infty}(G * F)^{n*} * G(t) \\ \sum_{n=0}^{\infty}(F * G)^{n*} * F(t) & \sum_{n=1}^{\infty}(F * G)^{n*}(t) \end{bmatrix}, \tag{7.4.4}$$

where Eq.(7.4.4) represents the mean number of visits to state j starting from state i, and where $i, j = 0, 1$. The Laplace-Stieltjes transform of $\mathbf{M}(t)$ is given by

$$\mathbf{M}^*(s) = \frac{1}{1 - G^*(s)F^*(s)} \begin{bmatrix} G^*(s)F^*(s) & G^*(s) \\ F^*(s) & F^*(s)G^*(s) \end{bmatrix}. \tag{7.4.5}$$

Noting that the process can move from one state to another alternately, we have the first passage time distribution:

$$G_{00}(t) = G_{11}(t) = G * F(t), \ G_{01}(t) = G(t), \ G_{10}(t) = F(t). \tag{7.4.6}$$

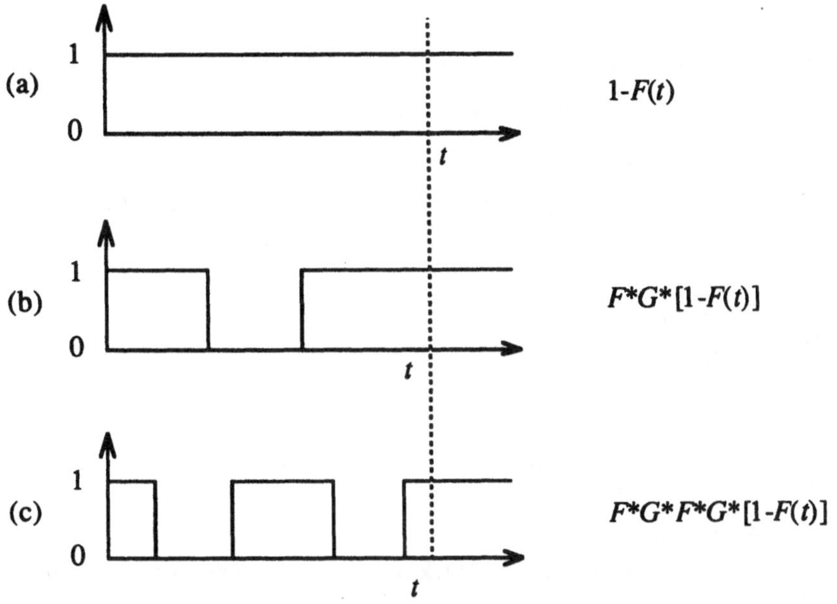

Fig. 7.4.1 An interpretation of the transition probability $P_{11}(t)$.

The transition probabilities are given by

$$P_{00}(t) = [1 - G] * \sum_{n=0}^{\infty} (G * F)^{n*}(t), \tag{7.4.7}$$

$$P_{11}(t) = [1 - F] * \sum_{n=0}^{\infty} (F * G)^{n*}(t), \tag{7.4.8}$$

$$P_{01}(t) = G * [1 - F] * \sum_{n=0}^{\infty} (G * F)^{n*}(t), \tag{7.4.9}$$

$$P_{10}(t) = F * [1 - G] * \sum_{n=0}^{\infty} (F * G)^{n*}(t). \tag{7.4.10}$$

For instance, the sample functions for deriving $P_{11}(t)$ in Eq.(7.4.8) are shown in Fig. 7.4.1. That is, Fig. 7.4.1 shows that (a) the process stays in state 1 for an interval $[0, t]$, (b) the process is in state 1 at time t after visiting state 0 once, and (c) the process is in state 1 at time t after visiting state 0 twice, and so on.

Let us consider the limiting behavior of an alternating renewal process. Assume that the process is aperiodic. If $F(t)$ and $G(t)$ are lattice with the same period δ, then the process is periodic with period δ. Otherwise, the process is aperiodic. We assume the latter, i.e., that the process is aperiodic. Noting that the embedded Markov chain is given in Eq.(7.4.2), we have the following stationary distribution of the embedded Markov chain:

$$\pi_0 = \pi_1 = \frac{1}{2}. \tag{7.4.11}$$

The unconditional means for states 1 and 0 are defined by

$$\int_0^\infty t\,dF(t) = \frac{1}{\lambda}, \quad \int_0^\infty t\,dG(t) = \frac{1}{\mu}, \tag{7.4.12}$$

respectively. Then the limiting probabilities of the alternating renewal process are

$$\rho_0 = \frac{\frac{1}{2}\cdot\frac{1}{\mu}}{\frac{1}{2}\cdot\frac{1}{\lambda}+\frac{1}{2}\cdot\frac{1}{\mu}} = \frac{\lambda}{\lambda+\mu}, \tag{7.4.13}$$

$$\rho_1 = \frac{\frac{1}{2}\cdot\frac{1}{\lambda}}{\frac{1}{2}\cdot\frac{1}{\lambda}+\frac{1}{2}\cdot\frac{1}{\mu}} = \frac{\mu}{\lambda+\mu}. \tag{7.4.14}$$

We finally derive the stationary Markov renewal process for the alternating renewal process:

$$P\{Z(0) = j\} = \rho_j \qquad (j = 0, 1), \tag{7.4.15}$$

$$P\{X(1) = 1,\ T_1 - T_0 \le t \mid X(0) = 0\}$$
$$= \mu \int_0^t [1 - G(y)]dy = G_e(t), \tag{7.4.16}$$

$$P\{X(1) = 0,\ T_1 - T_0 \le t \mid X(0) = 1\}$$
$$= \lambda \int_0^t [1 - F(y)]dy = F_e(t), \tag{7.4.17}$$

and

$$P\{X(n+1) = j,\ T_{n+1} - T_n \le t \mid X(n) = i\} = Q_{ij}(t)$$
$$(n = 1, 2, \cdots). \tag{7.4.18}$$

Then, the stationary process is constructed. For instance, if the process is in state 0 at time t given that it started from states 0 and 1, we have

$$P_{00}(t) = 1 - G_e(t) + \sum_{n=0}^{\infty} G_e * (F * G)^{n*} * F * [1 - G](t), \tag{7.4.19}$$

$$P_{10}(t) = \sum_{n=0}^{\infty} F_e * (G * F)^{n*} * [1 - G](t), \tag{7.4.20}$$

respectively. The Laplace-Stieltjes transforms of $F_e(t)$, $G_e(t)$, $F(t)$, and $G(t)$ are defined by $F_e^*(s)$, $G_e^*(s)$, $F^*(s)$, and $G^*(s)$, respectively, where

$$F_e^*(s) = \frac{\lambda[1 - F^*(s)]}{s}, \quad G_e^*(s) = \frac{\mu[1 - G^*(s)]}{s}. \tag{7.4.21}$$

Noting that the stationary process starts from states 0 and 1 with probabilities ρ_0 and ρ_1, respectively, we have

$$\rho_0 P_{00}(t) + \rho_1 P_{10}(t) = \frac{\lambda}{\lambda + \mu}, \tag{7.4.22}$$

since the Laplace-Stieltjes transform of the left-hand side of Eq.(7.4.22) is given by

$$\begin{aligned} &\rho_0 P_{00}^*(s) + \rho_1 P_{10}^*(s) \\ &= \frac{\lambda}{\lambda + \mu} \left\{ 1 - G_e^*(s) + \frac{G_e^*(s) F^*(s)[1 - G^*(s)]}{1 - F^*(s) G^*(s)} \right\} + \frac{\mu}{\lambda + \mu} \frac{F_e^*(s)[1 - G^*(s)]}{1 - F^*(s) G^*(s)} \\ &= \frac{\lambda}{\lambda + \mu}, \end{aligned} \tag{7.4.23}$$

which implies Eq.(7.4.22). Similarly, we can show that

$$\rho_0 P_{01}(t) + \rho_1 P_{11}(t) = \frac{\mu}{\lambda + \mu}. \tag{7.4.24}$$

Equations (7.4.22) and (7.4.24) show that the stationary distribution for states 0 and 1 is independent of time t and remains in the initial distribution.

Example 7.4.1 Assume that $F(t) = 1 - e^{-\lambda t}$ and $G(t) = 1 - e^{-\mu t}$ for an alternating renewal process. Then we have

$$F(t) = F_e(t) = 1 - e^{-\lambda t}, \quad G(t) = G_e(t) = 1 - e^{-\mu t},$$

and

$$\rho_0 P_{00}(t) + \rho_1 P_{10}(t) = \frac{\lambda}{\lambda + \mu},$$

$$\rho_0 P_{01}(t) + \rho_1 P_{11}(t) = \frac{\mu}{\lambda + \mu},$$

which have been given in Example 6.4.5 for a continuous-time Markov chain, where states 0 and 1 have been changed with one another. Note that the exponential distribution has the memoryless property which implies $F(t) = F_e(t)$ and $G(t) = G_e(t)$.

7.5 Problems 7

7.1 Consider a Markov renewal process $\{\mathbf{N}(t),\ t \geq 0\}$ with the semi-Markov kernel:

$$\mathbf{Q}(t) = \begin{matrix} 0 \\ 1 \end{matrix} \begin{bmatrix} 0.4(1-e^{-\lambda t}) & 0.6(1-e^{-\lambda t}) \\ 1-e^{-\mu t} & 0 \end{bmatrix}.$$

(i) Derive the transition probability matrix for the embedded Markov chain.

(ii) Compute the stationary probabilities ρ_j $(j = 0, 1)$ in Theorem 7.3.1.

7.2 (*Type I Counter*) Particles arrive at a particle counter with a Poisson process having parameter λ. An arriving particle finding the counter free gets *registered* and locks it for a random variable with distribution $G(t)$ $(t \geq 0)$ having mean $1/\mu$. Let states 0 and 1 denote that the counter is free and is locked, respectively.

(i) Derive the semi-Markov kernel $\mathbf{Q}(t)$.

(ii) Compute the stationary probabilities ρ_j $(j = 0, 1)$ for the Type I Counter.

7.3 (*Example 6.5.1*) Consider the job shop model in Example 6.5.1. Formulate this model in terms of a Markov renewal process.

(i) Show that the semi-Markov kernel is given by

$$\mathbf{Q}(t) = \begin{matrix} 0 \\ 1 \\ 2 \end{matrix} \begin{bmatrix} 0 & 1-e^{-\lambda t} & 0 \\ 0 & 0 & 1-e^{-\mu_1 t} \\ 1-e^{-\mu_2 t} & 0 & 0 \end{bmatrix}.$$

(ii) Derive the stationary distribution vector $[\pi_0 \quad \pi_1 \quad \pi_2]$ for the embedded Markov chain $\mathbf{Q}(\infty)$. (Note that the embedded Markov chain is irreducible, but is not aperiodic).

(iii) Computing the unconditional means ξ_j and using π_j $(j = 0, 1, 2)$ in (ii), derive the stationary probabilities ρ_j $(j = 0, 1, 2)$ and compare the results in Example 6.5.1.

7.4 (*Continuation*) Using the Laplace-Stieltjes transforms in Problem 7.3, determine the following:

(i) Derive the matrix Laplace-Stieltjes transform $\mathbf{Q}^*(s)$ of the semi-Markov kernel $\mathbf{Q}(t)$.

(ii) Derive $\mathbf{M}^*(s) = [\mathbf{I} - \mathbf{Q}^*(s)]^{-1} - \mathbf{I}$.

(iii) Derive $G^*_{ij}(s)$ for $i, j = 0, 1, 2$.

(iv) Derive $P^*_{ij}(s)$ for $i, j = 0, 1, 2$.

(v) Derive $P_{0j}(t)$ by inverting $P^*_{0j}(s)$ for $j = 0, 1, 2$.

7.5 Consider an alternating renewal process $\{\mathbf{N}(t),\ t \geq 0\}$ with the following semi-Markov kernel:

$$\mathbf{Q}(t) = \begin{matrix} 0 \\ 1 \end{matrix} \begin{bmatrix} 0 & G(t) \\ F(t) & 0 \end{bmatrix}.$$

(i) Verify that, using the renewal-theoretic arguments, the Markov renewal functions $M_{ij}(t)$ $(i, j = 0, 1)$ satisfy the following equations:

$$M_{00}(t) = G * M_{10}(t),$$

$$M_{01}(t) = G(t) + G * M_{11}(t),$$

$$M_{10}(t) = F(t) + F * M_{00}(t),$$

$$M_{11}(t) = F * M_{01}(t).$$

(ii) Introducing the Laplace-Stieltjes transforms $F^*(s)$, $G^*(s)$, and $M^*_{ij}(s)$, solve $M^*_{ij}(s)$ for $i, j = 0, 1$.

(iii) Derive $G^*_{ij}(s)$ and $P^*_{ij}(s)$ for $i, j = 0, 1$.

7.6 For a stationary alternating renewal process $\{\mathbf{N}(t), t \geq 0\}$ (see Section 7.4), verify Eq.(7.4.23), i.e.,

$$\rho_0 P^*_{01}(s) + \rho_1 P^*_{11}(s) = \frac{\mu}{\lambda + \mu}.$$

7.7 Consider a repairman problem of two machines (ref. Section 9.4.1). Once a machine breaks down, the machine is repaired immediately or waits to be repaired if the repairmen are busy. We assume that the failure and repair rates are constant, λ and μ, respectively, for each machine. Let j be the number of failed machines, where $j = 0, 1, 2$.

(i) For the repairman problem above with two repairmen, show that the problem is formulated by a Markov renewal process with the following semi-Markov kernel:

$$\mathbf{Q}(t) = \begin{matrix} 0 \\ 1 \\ 2 \end{matrix} \begin{bmatrix} 0 & 1 - e^{-2\lambda t} & 0 \\ \frac{\lambda}{\lambda+\mu}\left(1 - e^{-(\lambda+\mu)t}\right) & 0 & \frac{\mu}{\lambda+\mu}\left(1 - e^{-(\lambda+\mu)t}\right) \\ 0 & 1 - e^{-2\mu t} & 0 \end{bmatrix}.$$

(ii) For the repairman problem above with a repairman, show that the problem is formulated by a Markov renewal process, and derive the semi-Markov kernel.

Chapter 8

Reliability Models

8.1 Introduction

In the preceding chapters we have introduced several stochastic processes and developed their properties. In this chapter we discuss reliability models by using the results of the preceding chapters. There are fruitful applications of stochastic processes in reliability models.

In Section 8.2, we introduce the concept of the so-called "failure rate" for the lifetime distribution of an item, where "item" refers to any product such as a system, material, or part. Failure rate plays an important role in the following sections.

In Section 8.3, we develop the availability theory for a one-unit system which alternately assumes "up" and "down" states. Several "dependability" measures for such a system are introduced and derived under generalized assumptions. Analytical results are based on the direct applications of the alternating renewal process discussed in Section 7.4.

In Section 8.4, we discuss replacement models. Two replacement models, i.e., age and block replacement models, are introduced and developed. The monotonically increasing property of the failure rate implies an optimal replacement time under suitable assumptions.

In Section 8.5, we develop ordering models, which are an extension of replacement models. The monotonically increasing property of the failure rate again implies an optimal ordering time under suitable assumptions.

8.2 Lifetime Distributions and Failure Rates

We are interested in stochastic models in reliability theory. First of all, we introduce the following basic terms from MIL-STD 721 B (Military Standard, Department of Defense, U.S.A.).

Definition 8.2.1 *Item:* A non-specific term used to denote any product, including systems, materials, parts, subassemblies, sets, accessories, etc.

In this chapter we use systems and items interchangeably. In particular, if we develop stochastic models in reliability theory, we prefer system to item. However, in this section we are concerned with the lifetime distributions of an "item" which includes systems.

Definition 8.2.2 *Failure:* The event, or inoperable state, in which any item or part of an item does not, or would not perform as previously specified.

Our interest is to formulate a stochastic model by using the lifetime distribution of an item which can fail unpredictably.

Definition 8.2.3 *Random Failure:* Failure whose occurrence is predictable only in a probabilistic or statistical sense. This applies to all distributions.

Let X denote the *lifetime* of a system or an item subject to random failure, which is, of course, a random variable. The distribution of lifetime to failure is given by

$$F(t) = P\{X \leq t\} \qquad (t \geq 0). \tag{8.2.1}$$

In this chapter we assume that the failure law of an item is known, i.e., the lifetime distribution $F(t)$ in Eq.(8.2.1) is known. The survival probability of X is given by

$$\overline{F}(t) = 1 - F(t) = P\{X > t\} \qquad (t \geq 0), \tag{8.2.2}$$

which is the probability that the item survives up to time t and is called the *reliability function* of the item. We abbreviate the subscript X of $F(t)$ throughout this chapter.

As seen in Chapter 2, we have introduced the discrete and continuous random variables in general. We first assume that the random variable X is *continuous.* The probability density of the random variable X is assumed to exist and is given by

$$f(t) = \frac{dF(t)}{dt} \qquad (t \geq 0). \tag{8.2.3}$$

The *failure rate* or *hazard rate* is defined by

$$r(t) = \frac{f(t)}{\overline{F}(t)} \qquad (t \geq 0), \tag{8.2.4}$$

when $\overline{F}(t) > 0$. Noting that

$$r(t)dt = P\{t < X \leq t + dt \mid X > t\}, \tag{8.2.5}$$

we can interpret that $r(t)dt$ denotes the conditional probability that the item will fail during the time interval $(t, t + dt]$ given that it survives up to time t. Assuming $F(0) = 0$ (i.e., $\overline{F}(0) = 1$), we have

$$\overline{F}(t) = \exp\left[-\int_0^t r(x)dx\right], \tag{8.2.6}$$

and

$$f(t) = r(t)\exp\left[-\int_0^t r(x)dx\right]. \tag{8.2.7}$$

Equations (8.2.6) and (8.2.7) show that the reliability function (or distribution) and the probability density can be easily expressed in terms of its respective failure rate $r(t)$. That is, the failure rate $r(t)$ specifies its distribution, and vice versa. In particular, $\int_0^t r(x)\,dx$ is called the *cumulative hazard* or *hazard function.*

Definition 8.2.4 If the failure rate $r(t)$ is non-decreasing, the lifetime distribution $F(t)$ is called *IFR* (*Increasing Failure Rate*). If $r(t)$ is non-increasing, then $F(t)$ is called *DFR* (*Decreasing Failure Rate*).

It is sometimes true that an item whose lifetime distribution is IFR ages as time elapses. However, it is not easy to understand that an item whose lifetime distribution is DFR refreshes itself (i.e., its failure rate is non-increasing) as time elapses. The following phenomena can be found in human beings as well as in items. Figure 8.2.1 shows the typical failure rate curve. In general, the failure rate decreases (i.e., DFR) in the early phase (*infant phase*), is almost constant in the middle phase (*useful life phase*), and increases (i.e., IFR) in the final phase (*wearout phase*). Such a curve in Fig. 8.2.1 is called the *bath-tub curve* since it is quite similar to a bath-tub in shape.

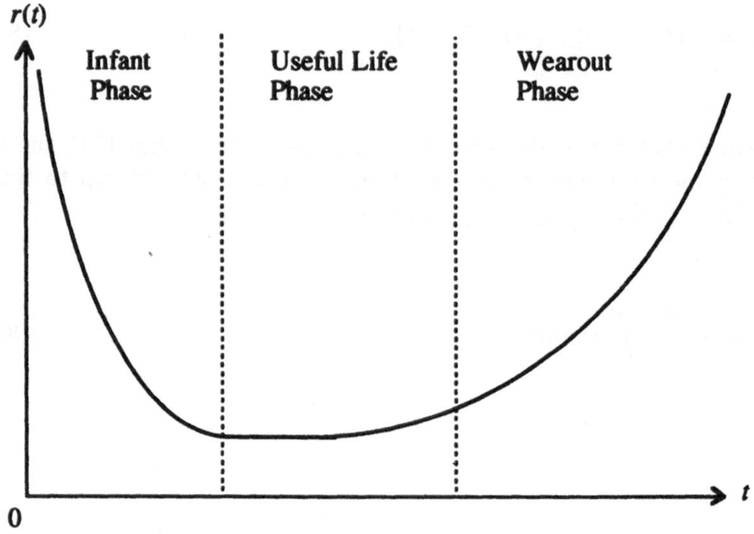

Fig. 8.2.1 A bath-tub curve of the failure rate $r(t)$.

We have discussed several continuous distributions in Section 2.4. We examine such distributions from the viewpoint of the failure rates.

(i) *Uniform Distribution* $X \sim U(a,b)$

The density and distribution of the uniform distribution in the interval $(a,b)(a < b)$ were given in Eqs.(2.4.1) and (2.4.2), respectively. The failure rate $r(t)$ is given by

$$r(t) = \frac{f(t)}{\overline{F}(t)} = \frac{1}{b-t} \qquad (a < t < b), \tag{8.2.8}$$

which is increasing. That is, the uniform distribution is IFR.

(ii) *Exponential Distribution* $X \sim EXP(\lambda)$

The failure rate $r(t)$ is given by

$$r(t) = \frac{f(t)}{\overline{F}(t)} = \frac{\lambda e^{-\lambda t}}{e^{-\lambda t}} = \lambda \qquad (t \geq 0), \tag{8.2.9}$$

which is constant. That is, an item whose lifetime distribution is exponential never ages as time elapses. The failure rate is constant if and only if $F(t)$ is exponentially distributed. Note also that the exponential distribution is both IFR and DFR because of Definition 8.2.4.

(iii) *Gamma Distribution* $X \sim GAM(\lambda, k)$

Noting that the reliability function $\overline{F}(t)$ is given by

$$\overline{F}(t) = \int_t^\infty \frac{\lambda(\lambda x)^{k-1}e^{-\lambda x}}{\Gamma(k)} dx \qquad (t \geq 0), \tag{8.2.10}$$

we have

$$\frac{1}{r(t)} = \frac{\overline{F}(t)}{f(t)} = \int_t^\infty \left(\frac{x}{t}\right)^{k-1} e^{-\lambda(x-t)}dx = \int_0^\infty \left(1+\frac{u}{t}\right)^{k-1} e^{-\lambda u}du, \tag{8.2.11}$$

where the change of the variable $u = x - t$ is made. We can show that $r(t)$ is decreasing (i.e., DFR) when $0 < k \leq 1$, and increasing (i.e., IFR) when $k \geq 1$. Of course, when $k = 1$, $r(t)$ is constant (i.e., the exponential distribution).

In particular, if k is a positive integer, we have

$$F(t) = 1 - \sum_{i=0}^{k-1} \frac{(\lambda t)^i}{i!} e^{-\lambda t}, \tag{8.2.12}$$

(see Theorem 3.3.2). The failure rate can be expressed by

$$r(t) = \frac{\dfrac{\lambda(\lambda t)^{k-1}e^{-\lambda t}}{(k-1)!}}{\displaystyle\sum_{i=0}^{k-1} \frac{(\lambda t)^i}{i!} e^{-\lambda t}}. \tag{8.2.13}$$

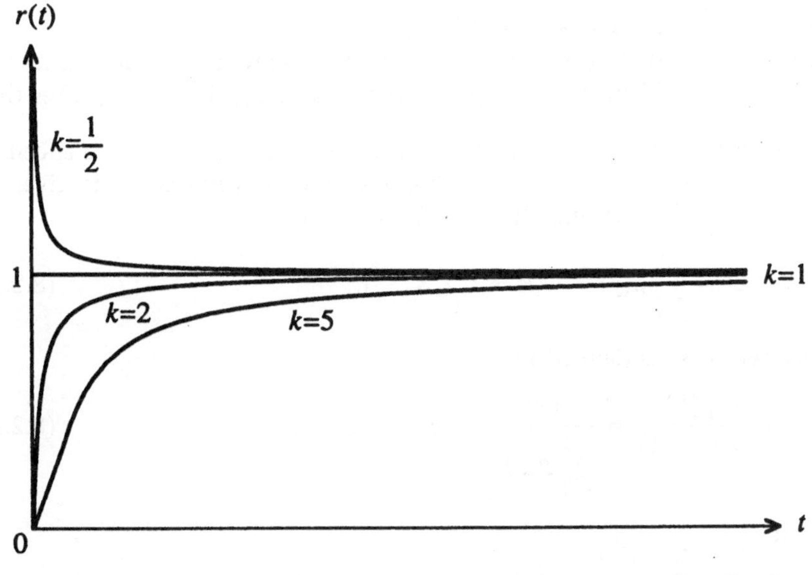

Fig. 8.2.2 The failure rate curves of the gamma distribution $X \sim GAM(\lambda, k)$ for $\lambda = 1$ and $k = 1/2, 1, 2, 5$.

Figure 8.2.2 shows the failure rate curves of the gamma distribution $X \sim GAM(\lambda, k)$ for $\lambda = 1$ and $k = 1/2, 1, 2, 5$.

(iv) *Weibull Distribution* $X \sim WEI(\alpha, \beta)$

The failure rate is given by

$$r(t) = \beta\alpha^\beta t^{\beta-1} \qquad (t \geq 0). \tag{8.2.14}$$

When $0 < \beta \leq 1$, the lifetime distribution is DFR, and when $\beta \geq 1$, it is IFR. Of course, when $\beta = 1$, the lifetime distribution is exponential.

(v) *Normal Distribution* $X \sim N(\mu, \sigma^2)$

The domain of the normal distribution is $(-\infty, \infty)$. If we apply the normal distribution to the lifetime distribution, the domain has to be $[0, \infty)$. That is, we introduce the following ***truncated normal distribution*** whose density is given by

$$f(t) = \frac{1}{a\sqrt{2\pi}\sigma} \exp\left[-\frac{(t-\mu)^2}{2\sigma^2}\right] \quad (t \geq 0,\ -\infty < \mu < \infty,\ \sigma > 0), \tag{8.2.15}$$

where

$$a = \int_0^\infty \frac{1}{\sqrt{2\pi}\sigma} \exp\left[-\frac{(x-\mu)^2}{2\sigma^2}\right] dx \tag{8.2.16}$$

is a normalizing constant and is introduced to hold the total probability $\int_0^\infty f(x)\,dx$ as a unity. For instance , if $\mu = 3\sigma, a \doteq 0.9987$. If $\mu = 2.5\sigma, a \doteq 0.9938$. In practice, if $\mu \geq 2.5\sigma$, we assume $a \doteq 1$. The truncated normal distribution is IFR. However, the proof needs much elaboration and we omit it.

(vi) *Lognormal Distribution* $X \sim LOG\,N(\mu, \sigma^2)$

The lognormal distribution can be adopted to maintenance. The failure rate is increasing (i.e., IFR) in the early phase and decreasing (i.e., DFR) after that.

We next discuss discrete distributions in general. Unless otherwise specified, the probability mass function of the discrete random variable X is given by $p(k)$ $(k = 1, 2, 3, \cdots)$. The reliability $\overline{F}(k)$ is given by

$$\overline{F}(k-1) = \sum_{j=k}^{\infty} p(j) \qquad (k = 1, 2, 3, \cdots). \tag{8.2.17}$$

The failure rate $r(k)$ is defined by

$$r(k) = \frac{p(k)}{\overline{F}(k-1)} = \frac{p(k)}{\sum_{j=k}^{\infty} p(j)} \qquad (k = 1, 2, \cdots). \tag{8.2.18}$$

Noting that

$$1 - r(k) = \frac{\sum_{j=k+1}^{\infty} p(j)}{\sum_{j=k}^{\infty} p(j)} \qquad (k = 1, 2, \cdots), \tag{8.2.19}$$

we have

$$\overline{F}(k-1) = \sum_{j=k}^{\infty} p(j) = \prod_{j=1}^{k-1} [1 - r(j)] \qquad (k = 1, 2, \cdots), \tag{8.2.20}$$

$$p(k) = r(k) \prod_{j=1}^{k-1} [1 - r(j)] \qquad (k = 1, 2, \cdots), \tag{8.2.21}$$

where we postulate $\prod_{j=1}^{k-1} = 1$ for $k = 1$.

Definition 8.2.5 If the failure rate $r(k)$ is non-decreasing in k, the discrete lifetime distribution $F(k)$ is called *IFR* (*Increasing Failure Rate*). If $r(k)$ is non-increasing in k, then $F(k)$ is called *DFR* (*Decreasing Failure Rate*).

To show that $F(k)$ is IFR or DFR, we should verify that

$$r(1) \leq r(2) \leq \cdots r(k) \leq \cdots,$$

or

$$r(1) \geq r(2) \geq \cdots r(k) \geq \cdots,$$

respectively.

We have discussed several discrete distributions in Section 2.3. We examine such distributions from the viewpoint of failure rates.

(i) *Uniform Distribution* $X \sim U(C + L, C + NL)$
The reliability function $\overline{F}(C + (k-1)L)$ is given by

$$\overline{F}(C + (k-1)L) = \frac{N - k + 1}{N} \qquad (k = 1, 2, \cdots, N). \tag{8.2.22}$$

The failure rate $r(k)$ is given by

$$r(C + kL) = \frac{1}{N - k + 1} \qquad (k = 1, 2, \cdots, N), \tag{8.2.23}$$

which is increasing in k (i.e., IFR).

We omit the discussions of (ii) Bernoulli distribution and (iii) binomial distribution since it is quite rare to apply them to the lifetime distribution. We proceed to the following:

(iv) *Geometric Distribution* $X \sim GEO(p)$
The reliability function $\overline{F}(k-1)$ is given by

$$\overline{F}(k-1) = \sum_{j=k}^{\infty} pq^{j-1} = q^{k-1} \qquad (k = 1, 2, \cdots). \tag{8.2.24}$$

The failure rate $r(k)$ is given by

$$r(k) = \frac{pq^{k-1}}{q^{k-1}} = p \qquad (k = 1, 2, \cdots), \tag{8.2.25}$$

which is constant in k. The failure rate is constant if and only if $X \sim GEO(p)$, i.e., the geometric distribution.

(v) *Negative Binomial Distribution* $X \sim NB(p, r)$

The negative binomial distribution is IFR when r is a positive integer greater than one, as is expected (see Example 2.3.2). That is, the negative binomial distribution corresponds to the gamma distribution.

(vi) *Poisson Distribution* $X \sim POI(\lambda)$

The Poisson distribution is IFR since

$$\begin{aligned} r(k+1) - r(k) &= \frac{p(k)}{\overline{F}(k-1)\overline{F}(k)} \left[\frac{\lambda}{k+1} \sum_{j=k}^{\infty} p(j) - \sum_{j=k+1}^{\infty} p(j) \right] \\ &= \frac{p(k)e^{-\lambda}}{\overline{F}(k-1)\overline{F}(k)} \sum_{j=k+1}^{\infty} \frac{\lambda^{j+1}}{j!} \left[\frac{1}{k+1} - \frac{1}{j+1} \right] > 0. \end{aligned} \tag{8.2.26}$$

8.3 Availability Theory

In the preceding section we introduced the basic terms from MIL-STD 721 B. We will once again use the MIL-STD 721 B for the following basic terms in reliability.

Definition 8.3.1 *Maintenance*: All actions necessary for retaining an item in, or restoring it to, a specified condition.

Definition 8.3.2 *Maintainability*: The measure of the ability of an item to be retained in or restored to specific condition when maintenance is performed by personnel having specified skill levels, using predescribed procedures and resources, at prespecified level of maintenance and repair.

Definition 8.3.3 *Corrective Maintenance*: All actions performed as a result of failure, to restore an item to a specified condition. Corrective maintenance can include any or all of the following steps: Localization, Isolation, Disassembly, Interchange, Reassembly, Alignments, and Checkout.

Definition 8.3.4 *Preventive Maintenance*: All actions performed in an attempt to retain an item in specified condition by providing systematic inspection, detection, and prevention of incipient failures.

Definition 8.3.5 *Scheduled Maintenance*: Preventive maintenance performed at prescribed points in the item's life.

In Section 8.4 we develop so-called "replacement models" in which a schedule of maintenance is performed to retain an item in specified condition by replacing it with a new item.

The following terms are from the IEC (International Electrotechnical Commission).

Definition 8.3.6 *Up Time*: The period of time during which an item performs its required function.

Definition 8.3.7 *Down Time*: The period of time during which an item is not in a condition to perform its required function.

Let X_i and Y_i $(i = 1, 2, \cdots)$ denote the lifetime (up time) and maintenance time (down time) of an item, where we assume that each item is identical and maintenance of the failed item can perform its required function. Consider a sample function shown in Fig. 8.3.1 that an item assumes up and down states alternately. We further assume that

$$F(t) = P\{X_i \leq t\} \quad (t \geq 0,\ i = 1, 2, \cdots), \tag{8.3.1}$$

$$G(t) = P\{Y_i \leq t\} \quad (t \geq 0,\ i = 1, 2, \cdots), \tag{8.3.2}$$

where all the random variables are assumed to be independent. Referring to Fig. 8.3.1, we are now in a position to define the following reliability and availability measures.

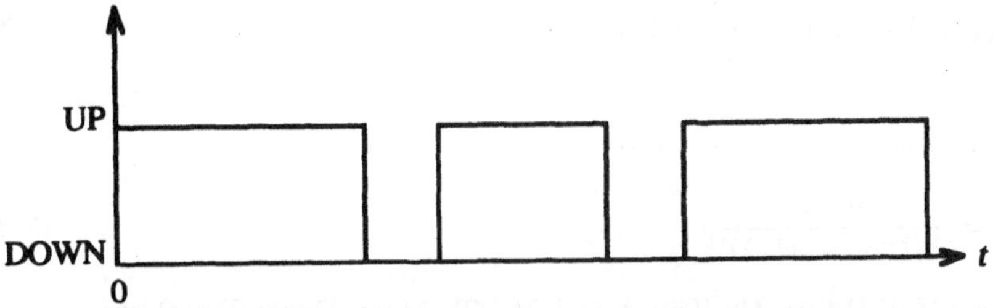

Fig. 8.3.1 A sample function of a process assuming up and down states alternately.

Definition 8.3.8 *Reliability* $R(t)$: The probability of an item performing its required function for the intended period of time $[0, t]$.

The *reliability* or *reliability function* is given by

$$R(t) = 1 - F(t) \qquad (t \geq 0) \tag{8.3.3}$$

which was shown in Eq.(8.2.2). The $MTTF$ (Mean Time To Failure) is given by

$$MTTF = \int_0^\infty t\, dF(t) = \int_0^\infty R(t)\, dt, \tag{8.3.4}$$

if $R(0) = 1$.

Definition 8.3.9 *Interval Reliability* $R(x,t)$: The probability that at a specified time t, an item is operating and will continue to operate for an interval of duration x.

Definition 8.3.10 *Limiting Interval Reliability* $R(x,\infty)$: It is defined by

$$R(x,\infty) = \lim_{t\to\infty} R(x,t). \tag{8.3.5}$$

Definition 8.3.11 *Availability* $A(t)$: The probability that an item is operating at a specified time t. This availability is referred to as *pointwise availability* or *instantaneous availability*.

Definition 8.3.12
(i) *Limiting Availability*:

$$A = \lim_{t\to\infty} A(t), \tag{8.3.6}$$

when it exists.
(ii) *Average Availability in* $[0,T]$:

$$A_{av}(T) = \frac{1}{T}\int_0^T A(t)\,dt. \tag{8.3.7}$$

(iii) *Limiting Average Availability*:

$$A_{av}(\infty) = \lim_{T\to\infty} \frac{1}{T}\int_0^T A(t)\,dt. \tag{8.3.8}$$

As will be shown later, we have

$$A = \frac{MUT}{MUT + MDT}, \tag{8.3.9}$$

where the MUT (Mean Up Time) and MDT (Mean Down Time) are

$$MUT = \int_0^\infty t\,dF(t), \qquad MDT = \int_0^\infty t\,dG(t), \tag{8.3.10}$$

respectively (see Fig. 8.3.1). The MUT and MDT are also referred to as the $MTBF$ (Mean Time Between Failures) and $MTBM$ (Mean Time Between Maintenances), respectively. It is well-known that the limiting average availability is the same as the limiting availability. That is, if $A = \lim_{t\to\infty} A(t)$ exists, then

$$A_{av}(\infty) = \lim_{T\to\infty} \frac{1}{T}\int_0^T A(t)\,dt = A. \tag{8.3.11}$$

We finally introduce the following measure.

Definition 8.3.13 The *joint availability* $A_{joint}(x,t)$ at t and $t+x$: The probability that the system is operating at t and again at $t+x$. The limiting joint availability is defined by

$$A_{joint}(x,\infty) = \lim_{t\to\infty} A_{joint}(x,t). \tag{8.3.12}$$

We develop the availability theory for a one-unit system which assumes up and down states alternately (see Fig. 8.3.1). We assume in Eqs.(8.3.1) and (8.3.2) that the failure (up) time distribution is a general $F(t)$ with finite mean $1/\lambda$ and the repair (down) time distribution is also a general $G(t)$ with finite mean $1/\mu$. We also assume that repair of the failed unit can reestablish its functioning perfectly. We derive the *dependability* measures defined above by analyzing the stochastic behavior for such a system.

As shown in Section 7.4, the model in which we are interested can be formulated by an alternating renewal process which is a Markov renewal process of a special type. Let $i = 0, 1$ denote the down and up states, respectively. Let $I_k(t)$ denote the binary indicator variable which assumes the value 1 (0) if the system is up (down) at time t, given that it was in state k at time $t = 0$, respectively, where $k = 0, 1$.

As shown in Fig. 8.3.1, the system repeats up and down states infinitely, and we define the Stieltjes convolution of $F(t)$ and $G(t)$:

$$H(t) = F * G(t), \tag{8.3.13}$$

which is the distribution of the sum of all up and down times. We define the following renewal function with the interarrival distribution $H(t)$:

$$M_H(t) = \sum_{n=1}^{\infty} H^{(n)}(t), \tag{8.3.14}$$

which will play a central role in later analysis. Note that from Theorem 4.3.2 (Elementary Renewal Theorem),

$$\lim_{t\to\infty} \frac{M_H(t)}{t} = \frac{1}{1/\lambda + 1/\mu} = \frac{\lambda\mu}{\lambda+\mu}. \tag{8.3.15}$$

Let $A_k(t) = P\{I_k(t) = 1\}$ $(k = 0, 1)$ denote the pointwise availability of the system given that it was in state k at time 0. Referring to the results in Section 7.4, we have

$$A_k(t) = P_{k1}(t) \qquad (k = 0, 1). \tag{8.3.16}$$

From Eqs.(7.4.8) and (7.4.9), we have

$$\begin{aligned} A_1(t) &= [1 - F] * \sum_{n=0}^{\infty} (F * G)^{n*}(t) \\ &= \overline{F}(t) + M_H * \overline{F}(t), \end{aligned} \tag{8.3.17}$$

$$A_0(t) = G * [1 - F] * \sum_{n=0}^{\infty} (F * G)^{n*}(t)$$
$$= G * \overline{F}(t) + G * M_H * \overline{F}(t), \tag{8.3.18}$$

where $\overline{F}(t) = 1 - F(t)$.

It is easy to show that

$$A = \lim_{t\to\infty} A_k(t) = \rho_1 = \frac{\mu}{\lambda + \mu} \qquad (k = 0, 1), \tag{8.3.19}$$

where ρ_1 was given in Eq.(7.4.14). The *limiting availability* A in Eq.(8.3.19) is independent of the initial state k. Noting that $\frac{1}{\lambda} = MUT$ and $\frac{1}{\mu} = MDT$, we have

$$A = \frac{\frac{1}{\lambda}}{\frac{1}{\lambda} + \frac{1}{\mu}} = \frac{MUT}{MUT + MDT}, \tag{8.3.20}$$

which was given in Eq.(8.3.9)

Example 8.3.1 If we assume that $F(t) = 1 - e^{-\lambda t}$ and $G(t) = 1 - e^{-\mu t}$, we have from Example 6.4.5 that

$$A_1(t) = \frac{\mu}{\lambda + \mu} + \frac{\lambda}{\lambda + \mu} e^{-(\lambda+\mu)t},$$

$$A_0(t) = \frac{\mu}{\lambda + \mu} - \frac{\mu}{\lambda + \mu} e^{-(\lambda+\mu)t},$$

The limiting availability is given by

$$A = \lim_{t\to\infty} A_k(t) = \frac{\mu}{\lambda + \mu} \qquad (k = 0, 1).$$

The average availability is given by

$$A_{av}(T) = \frac{1}{T}\int_0^T A_k(t)\, dt$$
$$= \begin{cases} \dfrac{\mu}{\lambda + \mu} + \dfrac{\lambda}{(\lambda+\mu)^2 T}[1 - e^{-(\lambda+\mu)T}] & (k = 1) \\ \dfrac{\mu}{\lambda + \mu} - \dfrac{\mu}{(\lambda+\mu)^2 T}[1 - e^{-(\lambda+\mu)T}] & (k = 0). \end{cases}$$

The limiting average availability is given by

$$A_{av}(\infty) = \lim_{T\to\infty} A_{av}(T) = A = \frac{\mu}{\lambda + \mu}.$$

We are interested in the interval reliability

$$R_k(x, t) = P\{I_k(u) = 1,\ u \in [t, t + x] \mid X(0) = k\}, \tag{8.3.21}$$

which is the survival probability that the system is up at time t and will continue to be up for an interval of duration x, where $X(0) = k$ denotes that the process is in state k at time 0. As in deriving the residual lifetime distribution in Eq.(4.3.37) for the renewal process, we have

$$R_1(x,t) = \overline{F}(t+x) + \int_0^t \overline{F}(t+x-u)\, dM_H(u), \tag{8.3.22}$$

and

$$R_0(x,t) = \int_0^t \overline{F}(t+x-u)\, dG(u) + \int_0^t \overline{F}(t+x-u)\, d[G * M_H(u)]. \tag{8.3.23}$$

Applying Theorem 4.3.6 (The Key Renewal Theorem), we have

$$R(x,\infty) = \lim_{t\to\infty} R_k(x,t) = \frac{\lambda\mu}{\lambda+\mu}\int_x^\infty \overline{F}(u)\, du \quad (k=0,1), \tag{8.3.24}$$

which is the limiting interval reliability. Note that Eq.(8.3.24) can be rewritten as

$$R(x,\infty) = \frac{\mu}{\lambda+\mu}\left\{\lambda\int_x^\infty \overline{F}(u)\, du\right\}. \tag{8.3.25}$$

Let

$$\psi(t) = \lambda\overline{F}(t) = \frac{\overline{F}(t)}{1/\lambda}, \quad \Psi(t) = \int_0^t \psi(u)\, du \tag{8.3.26}$$

be the density and distribution of the asymptotic distribution of the excess time (cf. Eq. (4.3.38)). Then we can rewrite

$$R(x,\infty) = \frac{\mu}{\lambda+\mu}[1-\Psi(x)], \tag{8.3.27}$$

where the right-hand side of the above equation is the product of the limiting probability (availability) that the system is up in the steady-state and the limiting probability that it can survive an interval of duration at least x.

Let us introduce the random variable $\gamma_k(t)$, the up excess random variable to down at time t, given $I_k(t) = 1 \quad (k=0,1)$. It is easy to show that

$$P\{\gamma_k(t) > x\} = \frac{R_k(x,t)}{A_k(t)} \quad (k=0,1), \tag{8.3.28}$$

and the up excess time distribution is given by

$$P\{\gamma_k(t) \le x\} = 1 - \frac{R_k(x,t)}{A_k(t)} \quad (k=0,1). \tag{8.3.29}$$

We are also interested in the mean up excess time:

$$E[\gamma_k(t)] = \int_0^\infty \frac{R_k(x,t)}{A_k(t)}\, dx \quad (k=0,1). \tag{8.3.30}$$

In particular,

$$E[\gamma_1(t)] = \frac{1}{A_1(t)} \int_0^\infty [\overline{F}(t+x) + \overline{F} * M_H(t+x)]\, dx, \tag{8.3.31}$$

$$E[\gamma_0(t)] = \frac{1}{A_0(t)} \int_0^\infty \overline{F} * [G(t+x) + G * M_H(t+x)]\, dx. \tag{8.3.32}$$

Noting that $\lim_{t\to\infty} A_k(t) = A$ $(k = 0, 1)$ in Eq.(8.3.19) and $\lim_{t\to\infty} R_k(x,t) = A[1 - \Psi(x)]$, we have

$$\lim_{t\to\infty} P\{\gamma_k(t) \le x\} = \Psi(x) = \lambda \int_0^x \overline{F}(u)\, du. \tag{8.3.33}$$

Assuming $t \to \infty$ in $E[\gamma_k(t)]$ $(k = 0, 1)$ and applying Theorem 4.3.6 (The Key Renewal Theorem), we have

$$\lim_{t\to\infty} E[\gamma_k(t)] = \frac{\lambda}{2} \int_0^\infty t^2 dF(t) \qquad (k = 0, 1). \tag{8.3.34}$$

We are interested in the joint availability ${}_kA_{joint}(x,t) = P\{I_k(t) = 1, I_k(t + x) = 1\}$ $(k = 0, 1)$ at times t and $t+x$, i.e., the joint availability is the probability that the system is up at t and again up at $t + x$ starting from state k at time 0.

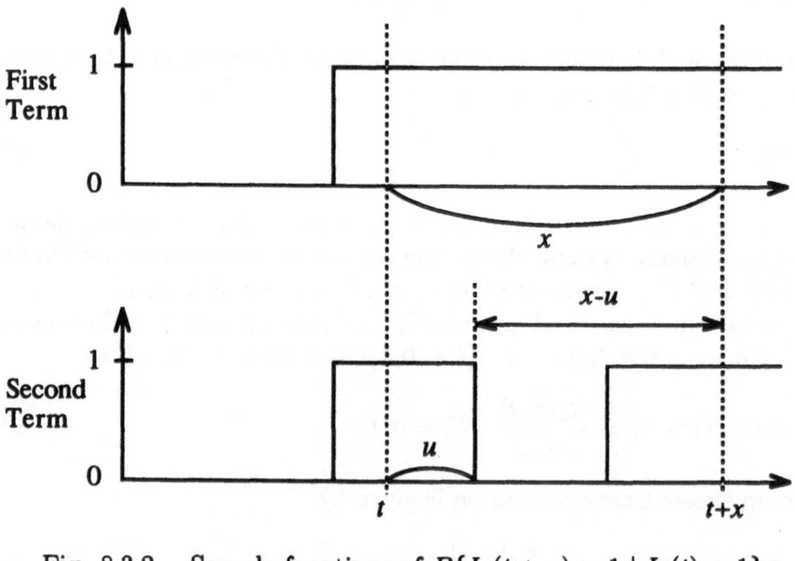

Fig. 8.3.2 Sample functions of $P\{I_k(t + x) = 1 \mid I_k(t) = 1\}$.

From renewal-theoretic argument, we have the following conditional probability

$$P\{I_k(t + x) = 1 \mid I_k(t) = 1\}$$

$$= \frac{R_k(x,t)}{A_k(t)} + \int_0^x A_0(x-u)dP\{\gamma_k(t) \le u\} \quad (k=0,1), \tag{8.3.35}$$

where the first term of the right-hand side denotes the probability that the system is up during $[t, t+x]$ and the second term the probability that the system is up at t, then goes down less than or equal to $t+u$, and follows the behavior of $A_0(t-u)$ during the remaining time $(t+u, t+x]$ (see Fig. 8.3.2). The joint availability is given by

$$\begin{aligned} &{}_kA_{joint}(x,t) \\ &\quad = P\{I_k(t+x)=1,\ I_k(t)=1\} \\ &\quad = P\{I_k(t+x)=1 \mid I_k(t)=1\}\, P\{I_k(t)=1\} \\ &\quad = R_k(x,t) + A_k(t)\int_0^x A_0(x-u)\, dP\{\gamma_k(t)\le u\} \qquad (k=0,1). \end{aligned} \tag{8.3.36}$$

It is easy to show that

$$\lim_{x\to\infty} {}_kA_{joint}(x,t) = \frac{\mu}{\lambda+\mu}A_k(t) \qquad (k=0,1), \tag{8.3.37}$$

and

$$\lim_{t\to\infty}\lim_{x\to\infty} {}_kA_{joint}(x,t) = \left(\frac{\mu}{\lambda+\mu}\right)^2 \qquad (k=0,1). \tag{8.3.38}$$

We now give the limiting joint availability:

$$\begin{aligned} A_{joint}(x,\infty) &= \lim_{t\to\infty} {}_kA_{joint}(x,t) \\ &= R(x,\infty) + \frac{\mu}{\lambda+\mu}\int_0^x [\lambda\overline{F}(u)]A_0(x-u)\, du \end{aligned} \tag{8.3.39}$$

which can be easily derived from renewal-theoretic arguments and the preceding results. Noting that $\Psi(x) = \int_0^x [\lambda\overline{F}(u)]\, du$, we have

$$A_{joint}(x,\infty) = \frac{\mu}{\lambda+\mu}\left[1 - \overline{A}_0 * \Psi(x)\right] = \frac{\mu}{\lambda+\mu}\left[1 - \Psi(x) + \Psi * A_0(x)\right], \tag{8.3.40}$$

where the bracket denotes the pointwise availability if the first up time distribution is $\Psi(t)$, and $\frac{\mu}{\lambda+\mu}$ is the limiting availability. That is, Eq.(8.3.40) denotes the probability that the system is up at some point (i.e., in the steady-state) and thereafter that the system is up at time x with the initial up time distribution $\Psi(t)$.

8.4 Replacement Models

In this section we develop several basic replacement models. As shown in Section 8.3, we have introduced two kinds of maintenance: Corrective maintenance and preventive maintenance. We are concerned with the latter whereby availability is increased and/or expected cost is reduced. Throughout this section we assume that the lifetime of an item is governed by the known continuous-time distribution $F(t)$ $(t \geq 0)$ with finite mean $1/\lambda$. We are interested in minimizing the expected cost rate (i.e., the expected cost per unit time in the steady-state) or maximizing the limiting availability.

8.4.1 Age Replacement Models

Consider a one-unit system whose lifetime distribution is given by $F(t)$ $(t \geq 0)$ with finite mean $1/\lambda$. We assume that there are an infinite number of identical units available for replacement. Consider a renewal process with interarrival time distribution $F(t)$ which corresponds to the process of a one-unit system with corrective maintenance. We define two scheduled maintenance policies.

Definition 8.4.1 *Age Replacement*: An item is replaced upon failure or at age t_0, whichever comes first.

Definition 8.4.2 *Block Replacement*: An item in operation is replaced upon failure and at times $T, 2T, 3T, \cdots$.

In this subsection we discuss age replacement models and in the next subsection we develop block replacement models.

Under an age replacement policy, an item is replaced upon failure or at age t_0, whichever comes first. Such a renewal process becomes a truncated renewal process with interarrival times $\{X_k, t_0\}$ $(k = 1, 2, \cdots)$. The stochastic behavior of such a process can be analyzed by using the results of the renewal process.

We are only interested in the expected cost per unit time in the steady-state which is referred to in the following as the *expected cost rate*. Let c_1 and c_2 denote the costs of failure (corrective maintenance) and scheduled replacement (preventive maintenance), respectively. It is natural to assume that

$$c_1 > c_2. \tag{8.4.1}$$

The expected cost rate can be given by the expected cost per cycle divided by the expected duration per cycle, where a *cycle* terminates whenever a renewal (an item is replaced upon failure or at age t_0, whichever comes first) takes place. The expected cost rate is given by

$$C(t_0) = \frac{c_1 P\{X_k \leq t_0\} + c_2 P\{X_k > t_0\}}{E[\min\{X_k, t_0\}]}, \tag{8.4.2}$$

since a cycle terminates at $\min\{X_k, t_0\}$ $(k = 1, 2, \cdots)$ and repeats itself again and again. Noting that

$$E[\min\{X_k, t_0\}] = \int_0^{t_0} t\, dF(t) + t_0\overline{F}(t_0) = \int_0^{t_0} \overline{F}(t)\, dt, \tag{8.4.3}$$

where $\overline{F}(t) = 1 - F(t) = P\{X_k > t\}$, we have

$$C(t_0) = \frac{c_1F(t_0) + c_2\overline{F}(t_0)}{\int_0^{t_0}\overline{F}(t)\, dt}. \tag{8.4.4}$$

Assuming that there exists the density $f(t)$ of $F(t)$, differentiating $C(t_0)$ with respect to t_0, and equating it to zero, we have

$$r(t_o)\int_0^{t_0}\overline{F}(t)\, dt - F(t_o) = \frac{c_2}{c_1 - c_2}, \tag{8.4.5}$$

where $r(t) = f(t)/\overline{F}(t)$, the failure rate. Equation (8.4.5) is a nonlinear equation with respect to $t_0 \geq 0$. Differentiating the left-hand side in Eq.(8.4.5), we have

$$r\prime(t_0)\int_0^{t_0}\overline{F}(t)dt,$$

which is non-negative (i.e., the left-hand side of Eq.(8.4.5) is a monotonically increasing function) if we assume $r'(t_0) \geq 0$, i.e., $F(t_0)$ is IFR. Noting that the left-hand side in Eq.(8.4.5) is increasing with the left-hand side $= 0$ at $t_0 = 0$ and the left-hand side tending to $r(\infty)/\lambda - 1$ as $t_0 \to \infty$ if $F(t_0)$ is IFR, we have the following theorem (see Fig. 8.4.1):

Theorem 8.4.1
(i) If $r(\infty) = \lim_{t\to\infty} r(t)$ exists and $r(\infty) > K$, there exists a finite t_0 such that $C(\infty) > C(t_0)$, where $K = \lambda c_1/(c_1 - c_2)$.

(ii) If $r(t)$ is continuous and monotonically increasing with $r(\infty) > K$, there exists a finite and unique t_0^* which satisfies the following equation:

$$r(t_0)\int_0^{t_0}\overline{F}(t)dt - F(t_0) = \frac{c_2}{c_1 - c_2}, \tag{8.4.6}$$

and

$$C(t_0^*) = (c_1 - c_2)r(t_0^*). \tag{8.4.7}$$

(iii) If $r(t)$ is continuous and monotonically increasing with $r(\infty) > K$, there exists a finite and unique $\overline{t_0}$ which satisfies $r(t_0) = K$, where $\overline{t_0} > t_0^*$ and $\overline{t_0}$ is an upper bound of the optimal t_0^*.

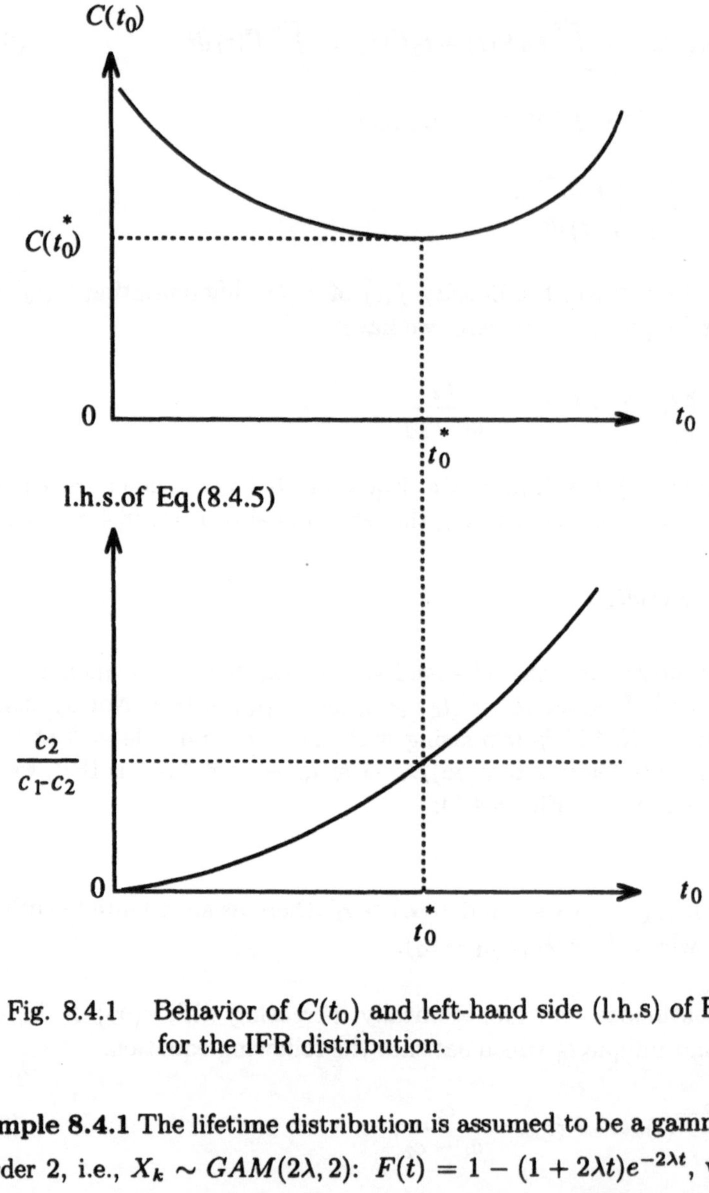

Fig. 8.4.1 Behavior of $C(t_0)$ and left-hand side (l.h.s) of Eq.(8.4.5) for the IFR distribution.

Example 8.4.1 The lifetime distribution is assumed to be a gamma distribution of order 2, i.e., $X_k \sim GAM(2\lambda, 2)$: $F(t) = 1 - (1 + 2\lambda t)e^{-2\lambda t}$, where $\dfrac{1}{\lambda}$ is the mean lifetime. The failure rate is given by $r(t) = \dfrac{4\lambda^2 t}{1 + 2\lambda t}$ and $r(\infty) = 2\lambda$. If $r(\infty) = 2\lambda > K = \dfrac{\lambda c_1}{c_1 - c_2}$, i.e., $\dfrac{c_1}{2} > c_2$, then there exists a finite and unique t_0^* satisfying Eq.(8.4.6), and $C(t_0^*) = (c_1 - c_2)r(t_0^*) = \dfrac{4\lambda^2 t_0^*(c_1 - c_2)}{1 + 2\lambda t_0^*}$. Otherwise, if $c_1/2 \le c_2$, the optimal policy is $t_0^* = \infty$, i.e., no scheduled maintenance (only corrective maintenance).

8.4.2 Block Replacement Models

For age replacement models, we have to observe the age of an item, which is sometimes difficult administratively. However, for block replacement models, we do not need to observe the age of an item, but replace at $T, 2T, 3T, \cdots$, which is easier to administer in general (see Definitions 8.4.1 and 8.4.2). In this subsection we develop the following three variations of block replacement models (see Fig. 8.4.2):

(I) A failed item is replaced instantaneously at failure.

(II) A failed item remains inoperable until the next scheduled replacement.

(III) A failed item undergoes minimal repair.

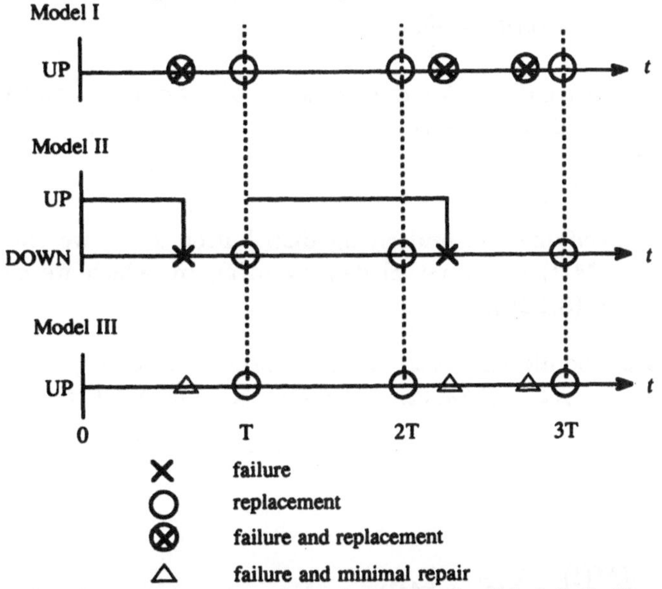

Fig. 8.4.2 Interpretations of the behavior for Models I, II, and III.

The models above are called *Models* I, II, and III, respectively. In general, Model I represents a typical block replacement model, and is well-known. Model III represents periodic replacement with minimal repair upon failure, where we assume that after each failure only *minimal repair* is performed so that the item's failure rate $r(t)$ is not disturbed. Model II is a simple variation of the block replacement model.

Model I

A failed item is replaced by a new item during the replacement interval T, and the scheduled replacement for the non-failed item is performed at $T, 2T, 3T, \cdots$.

Then the expected cost rate is

$$C_1(T) = \frac{c_1 M(T) + c_2}{T} = \frac{c_1 \int_0^T m(t)\,dt + c_2}{T}, \tag{8.4.8}$$

where

$$M(T) = \sum_{n=1}^{\infty} F^{(n)}(T) \tag{8.4.9}$$

is the renewal function with the underlying distribution $F(t)$, $m(t) = dM(t)/dt$ is the renewal density, c_1 is the cost of replacement for the failed item, and c_2 is the cost of scheduled replacement for the non-failed item. To minimize the expected cost rate in Eq.(8.4.8), we have the following equation by differentiating $C_1(T)$ and equating it to zero:

$$Tm(T) - \int_0^T m(t)\,dt = \frac{c_2}{c_1}. \tag{8.4.10}$$

This is a necessary condition for the existence of a finite T^*, and in this case, the resulting expected cost rate is

$$C_1(T^*) = c_1 m(T^*). \tag{8.4.11}$$

In general, in specifying the underlying distribution $F(t)$, we can obtain the renewal function $M(t)$ and renewal density $m(t)$, by which we can solve the nonlinear equation (8.4.10).

Example 8.4.2 The lifetime distribution is assumed to be a gamma a distribution of order 2, i.e., $X_k \sim GAM(2\lambda, 2)$. From Example 4.2.2, we have

$$M(t) = \lambda t - \frac{1}{4} + \frac{1}{4}e^{-4\lambda t},$$

and

$$m(t) = \frac{dM(t)}{dt} = \lambda(1 - e^{-4\lambda t}).$$

Eq.(8.4.10) becomes

$$\frac{1}{4}(1 - e^{-4\lambda T} - 4\lambda T e^{-4\lambda T}) = \frac{c_2}{c_1}.$$

If $\dfrac{c_2}{c_1} \geq \dfrac{1}{4}$, we should perform no scheduled replacement, i.e., an item is replaced only at failure. Otherwise, if $\dfrac{c_2}{c_1} < \dfrac{1}{4}$, there exists a finite and unique T^* which minimizes Eq.(8.4.8) and the resulting expected cost rate is

$$C_1(T^*) = c_1\lambda(1 - e^{-4\lambda T^*}).$$

Model II

For the first model we have assumed that a failed item is detected instantaneously just after failure. For Model II, we assume that failure is detected only at $T, 2T, 3T, \cdots$. An item is always replaced at $T, 2T, 3T, \cdots$, but is not replaced at the time of failure, and the item remains inoperable for the time duration from the occurrence of failure until its detection. The expected duration from the occurrence of failure until its detection per cycle is given by

$$\int_0^T (T-t)dF(t) = \int_0^T F(t)dt. \tag{8.4.12}$$

The expected cost rate is

$$C_2(T) = \frac{c_1 \int_0^T F(t)dt + c_2}{T}, \tag{8.4.13}$$

where c_1 is the cost of failure per unit time and c_2 is the cost of replacement for the non-failed item. Comparing Eq.(8.4.13) to Eq.(8.4.8), we have a similar expected cost rate, outside of the integrand. Similar discussions to those developed in Model I will now be presented.

Example 8.4.3 Referring to Eq.(8.4.10), we have

$$TF(T) - \int_0^T F(t)dt = \int_0^T [F(T) - F(t)]dt.$$

Assuming $T \to \infty$, we have

$$\lim_{T\to\infty} \int_0^T [F(T) - F(t)]dt = \int_0^\infty \bar{F}(t)dt = \frac{1}{\lambda}.$$

If $\frac{1}{\lambda} > \frac{c_2}{c_1}$, there exists an optimal T^* which is a unique and finite solution to

$$\int_0^T [F(T) - F(t)]dt = \frac{c_2}{c_1}.$$

If we assume the same distribution $F(t)$ as in Example 8.4.2, i.e., $X_k \sim GAM(2\lambda, 2)$, and $\frac{1}{\lambda} \le \frac{c_2}{c_1}$, we should perform no scheduled replacement. If $\frac{1}{\lambda} > \frac{c_2}{c_1}$, there exists a unique T^* satisfying

$$\frac{1}{2\lambda}[2 - e^{-2\lambda T}(2 + 4\lambda T + 4\lambda^2 T^2)] = \frac{c_2}{c_1},$$

and the resultant expected cost rate is

$$C_2(T^*) = c_1[1 - (1 + 2\lambda T^*)e^{-2\lambda T^*}].$$

Model III

We assume that minimal repair is performed when an item fails and the failure rate is not disturbed by each repair. If we consider a stochastic process $\{N(t), t \geq 0\}$ in which $N(t)$ represents the number of minimal repairs up to time t, the process $\{N(t), t \geq 0\}$ is governed by a nonhomogeneous Poisson process with mean value function

$$\Lambda(t) = \int_0^t r(x)dx, \tag{8.4.14}$$

which is the hazard function (see Section 8.2). Noting this fact, we have the following expected cost rate for Model III:

$$C_3(T) = \frac{c_1 \int_0^T r(t)dt + c_2}{T}, \tag{8.4.15}$$

where c_1 and c_2 are the costs of minimal repair for the failed item and of replacement for the non-failed item, respectively. Comparing Eq.(8.4.15) to Eq.(8.4.8), we also have the same expected cost rate except for the integrand. Similar discussions to those developed in Models I and II will be presented.

Example 8.4.4 If we assume the same distribution as in Example 8.4.2, i.e., $X_k \sim GAM(2\lambda, 2)$, we have

$$r(t) = \frac{4\lambda^2 t}{1 + 2\lambda t}.$$

There exists a unique T^* satisfying

$$\log(1 + 2\lambda T) - \frac{2\lambda T}{1 + 2\lambda T} = \frac{c_2}{c_1},$$

and the resultant expected cost rate is

$$C_3(T^*) = \frac{4c_1\lambda^2 T^*}{1 + 2\lambda T^*}.$$

We summarize three block replacement models: Models I, II, and III. The expected cost rate is given by

$$C_i(T) = \frac{c_1 \int_0^T \phi_i(t)dt + c_2}{T}, \tag{8.4.16}$$

where

$$\phi_i(t) = \begin{cases} m(t) & (i = 1; \textit{renewal density}) \\ F(t) & (i = 2; \textit{lifetime distribution}) \\ r(t) & (i = 3; \textit{failure rate}). \end{cases}$$

We also note the dimension of cost c_1 for each model: The cost c_1 for Models I and III is incurred for each failure, but the cost c_1 for Model II is incurred for each unit time that the item remains inoperable from the occurrence of failure until its detection.

To minimize $C_i(t)$ $(i = 1, 2, 3)$, we should solve the following equation:

$$\int_0^T [\phi_i(T) - \phi_i(t)]dt = \frac{c_2}{c_1}, \tag{8.4.17}$$

whose solution T^* satisfies a necessary condition of the minimum. If the solution T^* to Eq.(8.4.17) also satisfies a sufficient condition, the optimal expected cost rate is given by

$$C_i(T^*) = c_1\phi_i(T^*). \tag{8.4.18}$$

In practice, in most cases, by specifying $\phi_i(t)$, we can analytically demonstrate this necessary and sufficient condition, if it exists.

8.5 Ordering Models

In the replacement models developed in the preceding section, there are an unlimited number of items immediately available for replacement. However, this is not true in practice and here we assume that an item for any given replacement can be supplied only by an order with lead time. That is, only by making an order can we obtain an item for any given replacement after a lead time? The policy question that interests us is when we should order an item for each replacement. Thus, we call such a policy an ordering policy. Here we call such a model an *ordering model* in general.

Let us introduce the notation and assumptions used here. Assume that we can order an identical item for each replacement by an order. The random variable X denotes the lifetime of an operating item with an arbitrary distribution $F(t)$ $(t \geq 0)$ with finite mean $1/\lambda$. It is assumed that the failure of an operating item can be immediately detected. Let t_0 $(t_0 \geq 0)$ be the ordering time of an item, measured until the installation of the operating item. Let us next introduce the cost structures. A cost c_1 is incurred for each *expedited order* after the failure of an operating item. A cost c_2 is also incurred for each *regular order* before the failure of an operating item. It is plausible to assume $c_1 > c_2$, since an expedited order is much more expensive than a regular order. An ordered item can be obtained after a constant lead time $L \geq 0$, and is put in operation immediately or is in inventory (stock) up until the failure of an operating item. If an operating item fails with an ordered unit being not delivered, it cannot operate until the ordered unit is delivered. This incurs a cost k_1 per unit time for *shortage*. On the other hand, if an ordered unit is delivered before it is needed, then it is put into *inventory*. This incurs a constant cost k_2 per unit time for inventory.

Two ordering policies can be considered depending upon the situation of inventory of an ordered item: For *Model I*, an ordered item is put into inventory up until the failure of an operating unit that is operating, and, of course, an ordered item takes over operation immediately after delivery if the operating unit has been inoperable until then. For *Model II*, an ordered item takes over operation just after delivery, irrespective of the situation of the operating item.

8.5.1 Model I

We consider an infinite planning horizon. For an infinite planning horizon, it is appropriate to adopt an expected cost per unit time in the steady-state, i.e., the expected cost rate as an objective function.

Consider a cycle from the beginning of an item's operation until its replacement. Then, the expected cost of a cycle is given by the following three costs:
(i) The expected shortage cost is

$$k_1\left[\int_0^{t_0} L\,dF(t) + \int_{t_0}^{t_0+L}(t_0+L-t)\,dF(t)\right] = k_1\int_{t_0}^{t_0+L} F(t)\,dt, \quad (8.5.1)$$

since the shortage cost is proportional to the shortage time.
(ii) The expected inventory cost is

$$k_2\int_{t_0+L}^{\infty}(t-t_0-L)\,dF(t) = k_2\int_{t_0+L}^{\infty}\overline{F}(t)\,dt, \quad (8.5.2)$$

where $\overline{F}(t) = 1 - F(t)$.
(iii) The expected ordering cost is

$$c_1F(t_0) + c_2\overline{F}(t_0). \quad (8.5.3)$$

The mean time of a cycle is given by

$$\int_0^{t_0}(L+t)\,dF(t) + \int_{t_0}^{t_0+L}(t_0+L)\,dF(t) + \int_{t_0+L}^{\infty} t\,dF(t)$$

$$= \frac{1}{\lambda} + \int_{t_0}^{t_0+L} F(t)\,dt. \quad (8.5.4)$$

Thus, the total expected cost per one cycle is

$$C_1(t_0) = \frac{k_1\int_{t_0}^{t_0+L} F(t)\,dt + k_2\int_{t_0+L}^{\infty}\overline{F}(t)\,dt + c_1F(t_0) + c_2\overline{F}(t_0)}{\frac{1}{\lambda} + \int_{t_0}^{t_0+L} F(t)\,dt}, \quad (8.5.5)$$

which is the expected cost rate.

Our interest is to obtain the optimal ordering time t_0^* minimizing the expected cost $C_1(t_0)$ in Eq.(8.5.5) under the assumption that $c_1 > c_2$. It is assumed that there exists the density $f(t)$ of the lifetime distribution $F(t)$. We prove the following theorem which can be directly used for analysis.

Theorem 8.5.1 If the lifetime distribution $F(t)$ is IFR (DFR), then

$$F(x \mid t) = \frac{F(t+x) - F(t)}{\overline{F}(t)} \qquad (x > 0, t \geq 0) \tag{8.5.6}$$

is non-decreasing (non-increasing), respectively.

Proof If $F(t)$ is IFR, for $t_1 \leq t_2$, we have

$$r(t_1) \leq r(t_2), \tag{8.5.7}$$

which implies

$$\int_0^x r(t_1 + u)\, du \leq \int_0^x r(t_2 + u)\, du \tag{8.5.8}$$

and

$$\exp\left[-\int_{t_2}^{t_2+x} r(u)\, du\right] \leq \exp\left[-\int_{t_1}^{t_1+x} r(u)\, du\right]. \tag{8.5.9}$$

Using Eq.(8.2.6), we have

$$\frac{F(t_2 + x) - F(t_2)}{\overline{F}(t_2)} \geq \frac{F(t_1 + x) - F(t_1)}{\overline{F}(t_1)} \tag{8.5.10}$$

which proves that $F(x \mid t)$ is non-decreasing for $x > 0$ and $t \geq 0$. If $F(t)$ is DFR, the proof is just same except that the inequalities are inverted.

We assume that $C_1(\infty) < k_1$, i.e., $c_1 < k_1/\lambda$, since the expected cost of the system in which an order is made after the failure of an operating item would be less than that of the system which remains inoperative forever. Let

$$q_1(t) = F(L \mid t)\left[\frac{k_1}{\lambda} + k_2(L + \int_0^t \overline{F}(u)\, du) - c_1 F(t) - c_2 \overline{F}(t)\right]$$

$$+[(c_1 - c_2)r(t) - k_2]\left[\frac{1}{\lambda} + \int_t^{t+L} F(u)\, du\right], \tag{8.5.11}$$

for simplicity of equations. Then we have the following theorem:

Theorem 8.5.2 Assume that $c_1 < k_1/\lambda$:
(i) Suppose that $r(t)$ is monotonically increasing. If $q_1(0) < 0$ and $q_1(\infty) > 0$, then there exists a unique $t_0^* \in (0, \infty)$ which satisfies $q_1(t_0) = 0$. Otherwise, $t_0^* = \infty$ or 0 according to whether $q_1(\infty) \leq 0$ or $q_1(0) \geq 0$, respectively.
(ii) Suppose that $r(t)$ is non-increasing. Then, $t_0^* = \infty(0)$ if

$$\left(\frac{1}{\lambda} + L\right)\left(k_2 \int_L^\infty \overline{F}(t)\, dt - c_1 + c_2\right) \geq (<) \left(\frac{k_1}{\lambda} - c_1\right) \int_0^L \overline{F}(t)\, dt. \tag{8.5.12}$$

Proof Differentiating $C_1(t_0)$ with respect to t_0 and setting it equal to zero, we have $q_1(t_0) = 0$. Further, from the assumption that $c_1 < k_1/\lambda, q_1(t_0)$ is monotonically increasing (non-increasing) if $r(t_0)$ is monotonically increasing (non-increasing), respectively. First, suppose that $r(t_0)$ is monotonically increasing. If $q_1(0) < 0$ and $q_1(\infty) > 0$, then from the monotonicity and continuity of $q_1(t_0)$, there exists a unique $t_0^* \in (0, \infty)$ which satisfies $q_1(t_0) = 0$ and minimizes the expected cost $C_1(t_0)$. Further, it is easily shown that if $q_1(\infty) \leq 0$, then $t_0^* = \infty$, and if $q_1(0) \geq 0$, then $t_0^* = 0$.

Next, suppose that $r(t_0)$ is non-increasing. Then, $q_1(t_0)$ is also non-increasing. Thus, it is easily shown that $C_1(0)$ or $C_1(\infty)$ is not greater than $C_1(t_0)$ for any t_0. Therefore, we have $t_0^* = \infty$ if $C_1(\infty) \leq C_1(0)$, i.e.,

$$\left(\frac{1}{\lambda} + L\right)\left(k_2 \int_L^\infty \overline{F}(t)\,dt - c_1 + c_2\right) \geq (k_1/\lambda - c_1)\int_0^L \overline{F}(t)\,dt, \qquad (8.5.13)$$

and vice versa. We complete the proof.

In case of $q_1(0) < 0$ and $q_1(\infty) > 0$ of (i) in Theorem 8.5.2, the expected cost is given by

$$C_1(t_0^*) = k_1 + k_2 - \frac{k_2 - r(t_0^*)(c_1 - c_2)}{F(L \mid t_0^*)}. \qquad (8.5.14)$$

Further, the ordering policy when $t_0^* = \infty$ is that a spare item is ordered immediately after the failure of an operating item, and the ordering policy when $t_0^* = 0$ is that its order is made at the same time as the unit begins to operate.

8.5.2 Model II

In Model I, it has been assumed that a delivered unit is put into inventory if an operating item is still working. Here, Model II has the same assumptions as Model I except that an operating item is always replaced as soon as a spare item is delivered, even if it is operating. This model is appropriate in cases where the task of inventory is very difficult or there is no place to put a spare item in inventory. In this model, we do not need to consider inventory cost because of this assumption. The shortage cost is equal to Eq.(8.5.1). The total expected cost per one cycle is

$$C_2(t_0) = \frac{k_1 \displaystyle\int_{t_0}^{t_0+L} F(t)\,dt + c_1 F(t_0) + c_2 \overline{F}(t_0)}{L + \displaystyle\int_0^{t_0} \overline{F}(t)\,dt}. \qquad (8.5.15)$$

Our interest is to obtain the optimal t_0^* minimizing the expected cost $C_2(t_0)$ in Eq.(8.5.15) under the assumption that $c_1 > c_2$. Let

$$q_2(t) = [r(t) + b_1 F(L \mid t)]\left[L + \int_0^t \overline{F}(u)\,du\right] - F(t) - b_1 \int_t^{t+L} F(u)\,du,$$

(8.5.16)

where $b_1 = k_1/(c_1 - c_2)$ and $b_2 = c_2/(c_1 - c_2)$. Then, from discussion similar to Theorem 8.5.2, we obtain the following theorem without proof:

Theorem 8.5.3
(i) Suppose that $r(t)$ is monotonically increasing. If $q_2(0) < b_2$ and $q_2(\infty) > b_2$, there exists a unique $t_0^* \in (0, \infty)$ which satisfies $q_2(t_0) = b_2$. Otherwise, $t_0^* = \infty$ or 0 according to whether $q_2(\infty) \le b_2$ or $q_2(0) \ge b_2$, respectively.
(ii) Suppose that $r(t)$ is non-increasing. Then, $t_0^* = \infty(0)$ if

$$\left(\frac{1}{\lambda} + L\right)\left(k_1 \int_0^L \overline{F}(t)\,dt - c_1\right) \le (>) \left(\frac{k_1}{\lambda} - c_1\right) L. \tag{8.5.17}$$

Example 8.5.1 We assume that the lifetime distribution is a gamma distribution of order 2, i.e., $X_k \sim GAM(2\lambda, 2)$:

$$F(t) = 1 - (1 + 2\lambda t)e^{-2\lambda t}.$$

Then

$$r(t) = \frac{4\lambda^2 t}{1 + 2\lambda t}, \quad F(L \mid t) = 1 - e^{-2\lambda L} - \frac{2\lambda L e^{-2\lambda L}}{1 + 2\lambda t}.$$

Note that $r(t)$ is increasing with $r(0) = 0$ and $r(\infty) = 2\lambda$. From Theorem 8.5.3, if

$$\left(L + \frac{1}{\lambda}\right)\left[2\lambda + b_1(1 - e^{-2\lambda L})\right] > 1 + b_1 L + b_2$$

and

$$b_1 \left[\frac{1}{\lambda} - \left(\frac{1}{\lambda} + 2L + 2\lambda L^2\right) e^{-2\lambda L}\right] < b_2,$$

there exists a unique $t_0^* \in (0, \infty)$ which satisfies

$$A(1 + 2\lambda t) + Be^{-2\lambda t} = D,$$

where

$$A = \left(L + \frac{1}{\lambda}\right)[2\lambda + b_1(1 - e^{-2\lambda L})] - (1 + b_1 L + b_2),$$

$$B = 1 + b_1 L e^{-2\lambda L},$$

$$D = 2\lambda\left(L + \frac{1}{\lambda}\right)(1 + b_1 L e^{-2\lambda L}).$$

In this case, the expected cost rate (see Problem 8.12) is

$$C_2(t_0^*) = k_1(1 - e^{-2\lambda L}) + \frac{4\lambda^2 t_0^*(c_1 - c_2) - 2\lambda k_1 L e^{-2\lambda L}}{1 + 2\lambda t_0^*}.$$

8.6 Problems 8

8.1 (*Weibull Distribution*) Verify that the Weibull distribution is DFR when $0 < \beta \le 1$, and is IFR when $\beta \ge 1$.

8.2 Let $X_1, X_2, \cdots, X_n$ be independent lifetime distributions with the respective failure rates $r_1(t), r_2(t), \cdots, r_n(t)$. Verify that the failure rate of the random variable $\min(X_1, X_2, \cdots, X_n)$ is given by $\sum_{i=1}^{n} r_i(t)$.

8.3 Suppose that a series system of n items, which can fail if one of n items fails first, has independent exponential lifetimes with respective failure rates $\lambda_1, \lambda_2, \cdots, \lambda_n$.

(i) Compute the system reliability function.

(ii) Compute the mean and variance of the system lifetime.

8.4 Let $X_1, X_2, \cdots, X_n$ be independent and identical random variables with a distribution $F(t)$ $(t \ge 0)$. By ordering these n random variables from small to large, we can write $X_{1:n} \le X_{2:n} \le \cdots \le X_{n:n}$ which are called the *order statistics* from the distribution $F(t)$. Verify the following:

(i) $$P\{X_{i:n} \le t\} = \sum_{j=i}^{n} \binom{n}{j} [F(t)]^j [\bar{F}(t)]^{n-j}$$

$$= \frac{n!}{(i-1)!(n-i)!} \int_0^{F(t)} x^{i-1}(1-x)^{n-i}\, dx,$$

where $\overline{F}(t) = 1 - F(t)$.

(ii) $$E[X_{i:n}] = \int_0^{\infty} \sum_{j=0}^{i-1} \binom{n}{j} [F(t)]^j [\overline{F}(t)]^{n-j}\, dt.$$

8.5 (*Continuation*) Suppose that a system of n items can fail in the following ways:

(i) A *series system* which can fail if at least one of n items fails first.

(ii) A *parallel system* which can fail if at most n items fail.

(iii) A *k-out-of-n system* $(k \le n)$ which can fail if at least k out of n items fail.

Verify that the behavior of the three systems above can be described by the order statistics: (i) $X_{1:n}$, (ii) $X_{n:n}$, and (iii) $X_{k:n}$.

8.6 (*Example 8.3.1*) If we assume that $F(t) = 1 - e^{-\lambda t}$ and $G(t) = 1 - e^{-\mu t}$ ($t \geq 0$), compute the renewal function $M_H(t) = \sum_{n=1}^{\infty} H^{(n)}(t)$, where $H(t) = F * G(t)$.

8.7 (*Continuation*) Compute the interval reliability $R_k(x,t)$ and the limiting interval reliability $R_k(x,\infty)$ ($k = 0,1$) using the renewal function $M_H(t)$ in Problem 8.6.

8.8 Verify that, if the expected cost rate can be expressed by

$$C(t) = \frac{g(t)}{f(t)}$$

in replacement or ordering models, the optimal cost rate is given by

$$C(t) = \frac{g'(t)}{f'(t)}.$$

(Hint: To minimize $C(t)$, differentiate $C(t)$).

8.9 Verify that the necessary condition minimizing the expected cost rate $C(t_0)$ in an age replacement model is given by

$$r(t_0)\int_0^{t_0} \overline{F}(t)\,dt - F(t_0) = \frac{c_2}{c_1 - c_2}$$

by assuming that there exists the density $f(t)$ of $F(t)$.

8.10 (*Continuation*) Verify that, if $r(t) = f(t)/\overline{F}(t)$ is continuous and monotonically increasing with $r(\infty) > K = \lambda c_1/(c_1 - c_2)$, there exists a finite and unique t_0^* which satisfies the equation in Problem 8.9 and

$$C(t_0^*) = (c_1 - c_2)r(t_0^*).$$

8.11 (*Theorem 8.5.2*) Verify that, if $q_1(0) < 0$ and $q_1(\infty) > 0$, there exists a finite and unique t_0^* which satisfies $q_1(t_0) = 0$, and the optimal expected cost rate is

$$C_1(t_0^*) = k_1 + k_2 - \frac{k_2 - r(t_0^*)(c_1 - c_2)}{F(L|t_0^*)}.$$

8.12 (*Theorem 8.5.3*) Verify that, if $q_2(0) < b_2$ and $q_2(\infty) > b_2$, there exists a finite and unique t_0^* which satisfies $q_2(t_0) = b_2$, and the optimal expected cost rate is

$$C_2(t_0^*) = (c_1 - c_2)r(t_0^*) + k_1 F(L|t_0^*).$$

Chapter 9

Queueing Models

9.1 Introduction

We observe that people are waiting for service. Typical examples of such waiting lines are found at a supermarket, a restaurant, a bank, and so on. Such waiting lines are real waiting lines which we can observe directly. However, there are several waiting lines which we cannot observe directly. For instance, we often fail to complete a telephone call because the line is busy. An aircraft has to wait for its landing order because the runway is busy. A job or transaction has to wait its processing turn because the processor is busy.

Queueing models are stochastic models arising from waiting lines or queues. We reach fruitful results in queueing models by applying stochastic processes developed in the preceding chapters. In particular, many queueing models have been analyzed by applying birth and death processes. More elaborate queueing models can be analyzed by applying the semi-Markov processes from Chapter 7. In this chapter we develop queueing models by applying the birth and death processes from Chapter 6. However, we also discuss more generalized models by allowing arbitrary distribution for a few queueing models.

We first introduce the terms of queueing models. Potential *customers* arrive at a service facility expecting a *service*. If a customer arrives at the service facility in which a *server* is free, the customer can be served by the server immediately. Otherwise, the customer has to wait for the service, which constitutes a *queue*, if the server is busy with another customer. Let us consider a service facility. We can consider a single *channel* which corresponds to a single server, and multiple channels which correspond to multiple servers. Note that a multiple server queue model is a queueing model in which there is a single queue (or waiting line) with multiple servers.

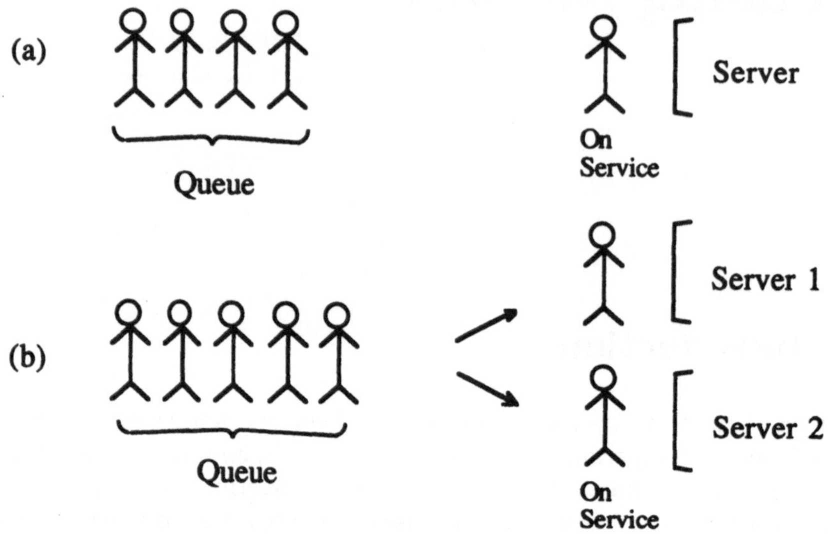

Fig. 9.1.1 (a) A single server queue, and (b) a multiple server queue.

We have introduced some terms such as customer, service, and server. As discussed earlier, we can consider many other queueing models by replacing relevant words for the terms above depending on the models we consider. For instance, in traffic theory or congestion theory the customer is replaced by the *call*, the service time by the ***holding time***, and the server by the ***telephone line***.

We can consider other queueing models. A Computer Reservation System (CRS) is one of the more sophisticated computer systems and is used for airline reservation systems, train reservation systems, and so on. A customer is a traveler who wants information, the service is an online inquiry for reserving vacant seat(s), and a server is an agent's terminal to the host computer.

Table 9.1.1 shows typical terms relevant to specific queueing models. Throughout this chapter we use the terms *customer*, *service*, *server*, and *service channel*. The reader should translate the relevant terms for the specific model as shown in Table 9.1.1.

To describe a queueing model, we have to specify the following six items:

(i) **Population**

The population (or source) of potential customers is either finite or infinite. Most queueing models are assumed to have an *infinite population*, where any

Table 9.1.1 Typical terms relevant to specific queueing models.

Queueing System	Traffic Theory	Repairman Problem	Airport	Computer Reservation	Information Processing
Customer	Call	Machine	Aircraft	Traveler	Transaction or Job
Customer's Arrival	Call's Arrival	Machine's Failure	Landing or Taking off	Traveler's Arrival	Transaction's or Job's Arrival
Server	Line	Repairman	Runway	Agent's Terminal	Communication Line plus Host Computer
Service Time	Holding Time	Repair Time	Landing or Taking off Time	Inquiry and Reservation Time	Processing Time

customer from an infinite population can arrive at a service facility. However, if we consider a repairman problem, such a problem has a *finite population* since the number of machines (which are potential customers) is finite.

(ii) **Interarrival Time Distribution**

Potential customers can arrive at a service facility expecting the service. Unless otherwise specified, we assume that customers can arrive at a service facility only one by one. Of course, there are situations where customers can arrive *en masse*. However, we never intend to discuss such an arrival pattern.

To describe the arrival pattern of a queue, we can consider interarrival time distributions for arriving customers. Several interarrival time distributions can be considered. We present the following interarrival time distributions.

Poisson Arrivals (Notation: M)

Potential customers arrive at a service facility with Poisson arrivals. That is, the interarrival time distributions for arriving customers obey the exponential distribution:

$$F(t) = 1 - e^{-\lambda t} \quad (t \geq 0), \tag{9.1.1}$$

where λ is the parameter of a Poisson (arrival) process and $1/\lambda$ is the mean arrival time for each customer. Of course, each interarrival time is independent (see Section 3.3). In particular, the parameter λ is called the *arrival rate* in queueing theory. As shown in Section 3.2, the Poisson process is used to describe "random" phenomenon and is called the "random" arrival process in queueing theory. For Poisson arrivals, we use the notation M which shows the *Markovian* (memoryless) property (see Kendall's notation below).

k-Erlang Distribution (Notation: E_k)

The interarrival time distributions for arriving customers obey a k-*Erlang distribution*:

$$F(t) = \int_0^t \frac{k\lambda(k\lambda x)^{k-1}e^{-k\lambda x}}{(k-1)!}\,dx$$

$$(t \geq 0,\ \lambda > 0,\ k: \text{a positive integer}), \qquad (9.1.2)$$

which is a gamma distribution of a special type (i.e., $T \sim GAM(k\lambda, k)$). The mean and variance of the k-Erlang distribution are given by

$$E[T] = \frac{1}{\lambda}\,,\ Var(T) = \frac{1}{k\lambda^2}\,. \qquad (9.1.3)$$

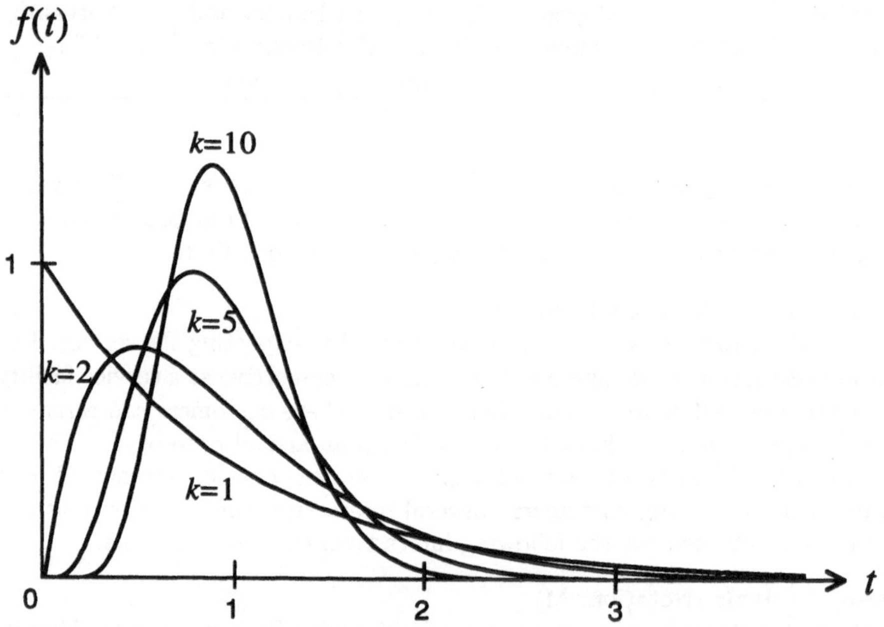

Fig. 9.1.2 The densities of the k-Erlang distribution for $\lambda = 1$, $k = 1, 2, 5, 10$.

Figure 9.1.2 shows the densities of the k-Erlang distribution for $\lambda = 1$, $k = 1, 2, 5, 10$. From Fig. 9.1.2, we can understand that $k = 1$ implies the exponential distribution and $k \to \infty$ implies a degenerate distribution at $t = 1/\lambda$ (see below).

Degenerate Distribution (Notation: D)

The interarrival time distributions for arriving customers are regular, i.e., each customer arrives regularly at $t = 1/\lambda$:

$$F(t) = \begin{cases} 0 & (t < 1/\lambda) \\ 1 & (t \geq 1/\lambda), \end{cases} \qquad (9.1.4)$$

where the mean and variance of the degenerate distribution are given by

$$E[T] = \frac{1}{\lambda}, \quad Var(T) = 0. \tag{9.1.5}$$

General Distribution (Notation: G or GI)

We can consider that the interarrival times for arriving customers are independent and identically distributed random variables with an arbitrary distribution $F(t)$ $(t \geq 0)$. We use the notation G or GI which shows the *general distribution* or *independent general distribution*, respectively. That is, the arriving customers obey a renewal process with interarrival time distribution $F(t)$.

We have introduced four arrival patterns. Throughout this chapter we develop the queueing theory with Poisson arrivals. However, many results have been obtained by assuming non-Poisson arrivals. Such results go beyond the level of this textbook.

(iii) **Service Time Distribution**

We can apply the same distributions discussed above (interarrival time distributions) for the service time distribution, where we introduce the *service rate* μ instead of the arrival rate λ. We note that Poisson arrivals mean that customers arrive one by one by following the Poisson process. However, if we consider the service time distribution as a Poisson process, the service time distribution obeys the exponential distribution for an arriving customer if there is at least one arriving customer. Thus, we call such a service time distribution an *exponential service.*

We mainly discuss queueing models with Poisson arrival and exponential service throughout this chapter.

(iv) **Maximum Queueing System Capacity**

We can consider two categories of maximum queueing system capacity: The maximum capacity is assumed to be *infinite*, i.e., all the arriving customers can be allowed to wait for service irrespective of the queueing length. Or, the maximum capacity is assumed to be *finite*, i.e., if a customer arrives when maximum capacity has been reached (i.e., no space to queue), the customer is turned away. In particular, if there is no queueing space, we speak of a "loss system," examples of which are queueing models in traffic theory (see Subsections 9.3.2 and 9.3.3).

(v) **Number of Service Channels**

We can consider a *single server queueing model* as one of the simplest queueing models. If there are more than one server, we call such a queueing model a *multiple server queueing model.* Of course, it is true that there is a single waiting line for both single and multiple server queueing models (see Fig. 9.1.1).

(vi) **Queueing Discipline**

There are customers who wait for service. The rules determining the order of service are called the *queue discipline.* The most popular and well-known queue discipline is *first come, first served* (FCFS). Throughout this chapter, we develop queueing models with FCFS. Of course, we can consider other queue discipline such as *last come, first served* (LCFS) and *random selection for service* (RSS). Most results in queueing models are the same irrespective of the queue discipline. However, some results, such as waiting time, are completely different for each kind of queue discipline.

Let us introduce *Kendall's notation* for queueing models:

$$\mathrm{A}/\mathrm{B}/c/K/m/\mathrm{Z}.$$

That is, A denotes the arrival pattern, B the service pattern, c the number of service channels, K the maximum queueing system capacity, m the size of the population, and Z the queue discipline. We can specify M, E_k, D, G, and GI for A and B. We note that the maximum queueing system capacity refers to the maximum numbers of customers in queue plus service channels.

For instance, an $\mathrm{M}/\mathrm{M}/1/\infty/\infty/\mathrm{FCFS}$ queue model is a model with Poisson arrival, exponential service, single server, infinite capacity, infinite population and first come, first served queue discipline. In practice, we use the shorthand notation, $\mathrm{M}/\mathrm{M}/1/\infty$ queue model, since the last two letters are trivial except in Section 9.4.

We mainly discuss M/M queueing models in this chapter. By assuming M/M queueing models, we can formulate such queueing models by using the birth and death processes from Chapter 6. Several results obtained in Chapter 6 can be directly used for purposes of analysis.

In queueing theory, the following queueing models have many interesting results:

- M/G/1 queueing models,
- GI/M/1 queueing models.

However, our intention is not to discuss such results in general. We just sketch some discussions of such queueing models.

In Section 9.2 we discuss M/M/1 queueing models by classifying maximum queueing system capacity as infinite or finite. Some basic and important results for M/M/1 queueing models are given.

In Section 9.3 we develop $\mathrm{M}/\mathrm{M}/c$ queueing models, i.e., multiple server queueing models with Poisson arrival and exponential service. We discuss three queueing models: an $\mathrm{M}/\mathrm{M}/c/\infty$ queueing model, an $\mathrm{M}/\mathrm{M}/c/c$ queueing model, and an $\mathrm{M}/\mathrm{M}/\infty/\infty$ queueing model.

In Section 9.4 we discuss M/M queueing models with a finite population. Typical examples of such queues are (machine) repairman problems. We analyze such queueing models by using the terms from repairman problems.

9.2 Single Server Queueing Models

Consider a single server queueing model with Poisson arrival and exponential service. We assume that the population of potential customers is infinite and queue discipline is FCFS.

Let $F(t) = 1 - e^{-\lambda t}$ be the interarrival time distribution for arriving customers. For a small $h > 0$, we have

$$F(h) = 1 - e^{-\lambda h} = \lambda h + o(h). \tag{9.2.1}$$

That is, the *arrival* (birth) *rate* for a birth and death process is given by

$$\lambda_k = \lambda \quad (k = 0, 1, 2, \cdots), \tag{9.2.2}$$

which is constant irrespective of varying k. Note the *mean interarrival time* is given by $1/\lambda$.

Let $G(t) = 1 - e^{-\mu t}$ be the service time distribution for arriving customers. Similar to Eq.(9.2.1), we have

$$G(h) = 1 - e^{-\mu h} = \mu h + o(h), \tag{9.2.3}$$

for a small $h > 0$, which implies the *service* (death) *rate* is

$$\mu_{k+1} = \mu \quad (k = 0, 1, 2, \cdots). \tag{9.2.4}$$

Note that the *mean service time* is given by $1/\mu$.

We discuss M/M/1 queueing models in which the arrival rate $\lambda_k = \lambda$ and service rate $\mu_{k+1} = \mu$ are constant, irrespective of $k = 0, 1, 2, \cdots$. We classify M/M/1 queueing models as examples of an M/M/1/∞ queueing model or an M/M/1/N queueing model.

9.2.1 M/M/1/∞ Queueing Models

As described above, we assume that

$$\lambda_k = \lambda, \quad \mu_{k+1} = \mu \qquad (k = 0, 1, 2, \cdots), \tag{9.2.5}$$

for an M/M/1/∞ queueing model. Referring to Theorem 6.4.2, we find that the following necessary and sufficient condition has to be satisfied

$$\sum_{j=0}^{\infty} \prod_{k=1}^{j} \frac{\lambda_{k-1}}{\mu_k} = \sum_{j=0}^{\infty} \left(\frac{\lambda}{\mu}\right)^j = \sum_{j=0}^{\infty} \rho^j < \infty, \tag{9.2.6}$$

in order for the limiting probabilities $p_j = \lim_{t\to\infty} P_{ij}(t)$ to exist, where we postulate $\prod_{k=1}^{j} \cdot = 1$ for $j = 0$. For Eq.(9.2.6) to hold, we assume that

$$\rho = \frac{\lambda}{\mu} < 1 \qquad \text{or} \qquad \lambda < \mu, \tag{9.2.7}$$

where

$$\rho = \frac{\lambda}{\mu} = \frac{\text{[arrival rate]}}{\text{[service rate]}} = \frac{1/\mu}{1/\lambda} = \frac{\text{[mean service time]}}{\text{[mean interarrival time]}} \tag{9.2.8}$$

is called the *traffic intensity* in queueing theory. It is plausible that the necessary and sufficient condition is $\rho < 1$, i.e., [mean interarrival time=$1/\lambda$]>[mean service time=$1/\mu$]. However, if $\rho \geq 1$, there exist no limiting probabilities p_j $(j = 0, 1, 2, \cdots)$, since the queue length becomes infinite.

Let $\{X(t), t \geq 0\}$ be a birth and death process describing an M/M/1/∞ queueing model. The behavior in the steady-state is denoted by $X(\infty) = X$. For $\rho < 1$, from Theorem 6.4.2 and Example 6.4.2, we have

$$p_j = (1-\rho)\rho^j \qquad (j = 0, 1, 2, \cdots), \tag{9.2.9}$$

which is the geometric distribution $X \sim GEO(1-\rho)$. Note that p_j means the limiting probability that there are j customers in the system, i.e., $(j-1)$ customers waiting for service and a customer being served in the steady-state. The mean and variance of X is given by

$$L = E[X] = \sum_{j=1}^{\infty} j(1-\rho)\rho^j = \frac{\rho}{1-\rho}, \tag{9.2.10}$$

$$Var(X) = E[X^2] - E[X]^2 = \frac{\rho}{(1-\rho)^2}. \tag{9.2.11}$$

That is, the mean and variance of the number of customers in the system are given by L and $Var(X)$, respectively.

Example 9.2.1 Consider the mean L and variance $Var(X)$ for varying ρ $(0 < \rho < 1)$. Table 9.2.1 shows the numerical examples for varying ρ. As shown in Eqs.(9.2.10) and (9.2.11), we see that L and $Var(X)$ tend to infinity as $\rho \to 1$. That is, the mean and variance rapidly increase as the traffic intensity ρ tends toward a unity.

Table 9.2.1 *L* and *Var(X)* for varying ρ.

ρ	L	$Var(X)$
$\frac{1}{4}$	$\frac{1}{3}$	$\frac{4}{9}$
$\frac{1}{2}$	1	2
$\frac{2}{3}$	2	6
$\frac{3}{4}$	3	12
$\frac{9}{10}$	9	90
$\frac{19}{20}$	19	380

Let L_q be the mean number of customers in a queue, i.e., the mean number of customers waiting for service excluding a customer being served. Noting that the real number of customers waiting for service is $j-1$ if j customers are in the system, we have

$$L_q = \sum_{j=1}^{\infty}(j-1)p_j = \sum_{j=1}^{\infty} jp_j - \sum_{j=1}^{\infty} p_j = L - \rho = \frac{\rho^2}{1-\rho}, \tag{9.2.12}$$

where

$$P\{\text{the server is free in the steady state}\} = p_0 = 1-\rho, \tag{9.2.13}$$

$$P\{\text{the server is busy in the steady state}\} = \sum_{j=1}^{\infty} p_j = \rho. \tag{9.2.14}$$

Referring to Eq.(9.2.14), we call ρ the *utilization factor* since $P\{$the server is busy$\}=\rho$.

A queueing process can be described by an alternating renewal process as in Section 7.4. That is, the process repeats the idle period and busy period alternately. Noting that the mean idle time is given by $1/\lambda$ and $P\{$the server is busy$\}=\rho$, we have

$$\frac{1/\lambda}{1/\lambda + [\text{average busy period}]} = 1-\rho, \tag{9.2.15}$$

which implies

$$[\text{average busy period}] = \frac{1}{\mu(1-\rho)}. \tag{9.2.16}$$

Consider the *waiting time* U for a customer who just joined the queue. Noting that the probability that there are n customers in the system is p_n, and the service time distribution for consecutive n customers obeys a gamma distribution $GAM(\mu, n)$, i.e., the n-fold convolution of the exponential service time:

$$dG_n(t) = \frac{\mu(\mu t)^{n-1}e^{-\mu t}}{(n-1)!}dt \qquad (t \geq 0), \tag{9.2.17}$$

we have the following probability, where U denotes the waiting time:

$$\begin{aligned} dP\{U \leq t\} &= \sum_{n=1}^{\infty} p_n dG_n(t) \\ &= \sum_{n=1}^{\infty}(1-\rho)\rho^n \frac{\mu(\mu t)^{n-1}e^{-\mu t}}{(n-1)!}dt \\ &= \lambda(1-\rho)e^{-\mu t}\sum_{n=1}^{\infty}\frac{(\lambda t)^{n-1}}{(n-1)!}dt \\ &= \lambda(1-\rho)e^{-(\mu-\lambda)t}dt. \end{aligned} \tag{9.2.18}$$

That is, $P\{U \leq t\}$ is a distribution of exponential type. Noting that the probability that there are no customers in the system is p_0 and that an arriving customer is served immediately, we have

$$\begin{aligned} P\{U \leq t\} &= P\{U = 0\} + P\{0 < U \leq t\} \\ &= 1 - \rho + \int_0^t \lambda(1-\rho)e^{-(\mu-\lambda)x}dx \\ &= 1 - \rho e^{-(\mu-\lambda)t}. \end{aligned} \tag{9.2.19}$$

The distribution $P\{U \leq t\}$ has a positive probability $1 - \rho$ at $t = 0$. The mean and variance of the waiting time are given by

$$W_q = E[U] = \int_0^\infty t dP\{U \leq t\} = \frac{\rho}{\mu(1-\rho)}, \tag{9.2.20}$$

$$Var(U) = \frac{\rho(2-\rho)}{\mu^2(1-\rho)^2}. \tag{9.2.21}$$

Let us consider the time spent in the system, which is the sum of the waiting time U and the service time V. Noting that the service time V is independent of the waiting time U and obeys the exponential distribution $P\{V \leq t\} = 1 - e^{-\mu t}$, we have the distribution of the time spent in the system as the following Stieltjes convolution:

$$\begin{aligned} P\{U + V \leq t\} &= \int_0^t P\{U \leq t - x\} dP\{V \leq x\} \\ &= \int_0^t [1 - \rho e^{-\mu(1-\rho)(t-x)}] \mu e^{-\mu x} dx \\ &= 1 - e^{-\mu(1-\rho)t}, \end{aligned} \tag{9.2.22}$$

which is the exponential distribution with parameter $\mu(1-\rho)$. The mean and variance of the time spent in the system are given by

$$W = E[U + V] = \frac{1}{\mu(1-\rho)}, \tag{9.2.23}$$

$$Var(U + V) = \frac{1}{\mu^2(1-\rho)^2}. \tag{9.2.24}$$

Referring to the mean number of customers in the system L in Eq.(9.2.10), the mean of the time spent in the system W in Eq.(9.2.23), the mean number of customers in the queue L_q in Eq.(9.2.12), and the mean waiting time W_q in Eq.(9.2.20), we have the following formulas:

$$L = \lambda W, \tag{9.2.25}$$

$$L_q = \lambda W_q, \tag{9.2.26}$$

which are called *Little's Formulas.* We have discussed Little's formulas for M/M/1 queueing models. However, Little's formulas have a wider scope of validity if one allows for the less stringent conditions of many queueing models. Little's formulas will be frequently discussed in the following sections.

Example 9.2.2 Word processing in a small office can be formulated in terms of an $M/M/1/\infty$ queueing model. Assume that the mean arrival time for word processing is 25 minutes and the mean service time for word processing is 15 minutes. Calculate the following:
(i) The probability that the word processor is busy.
(ii) The mean waiting time.
(iii) If the demand for word processing increases and the mean time spent in the system is over 45 minutes, we should introduce another word processor. How do we decide the critical mean arrival time?

Solution (i) Noting that $1/\lambda = 25$ min. and $1/\mu = 15$ min., we have the traffic intensity $\rho = \lambda/\mu = 3/5$. Thus, $P\{\text{busy}\} = \rho = 3/5$.
(ii) $W_q = \dfrac{\rho}{\mu(1-\rho)} = 22.5$ min.
(iii) Assuming that λ is unknown, we have the following inequality:

$$W = \frac{1}{\mu(1-\rho)} = \frac{1/\mu}{1-\lambda/\mu} = \frac{15}{1-15\lambda} \geq 45,$$

which implies $1/\lambda \leq 22.5$ min. That is, if the mean arrival time for word processing is less than or equal to 22.5 min., we should introduce another word processor.

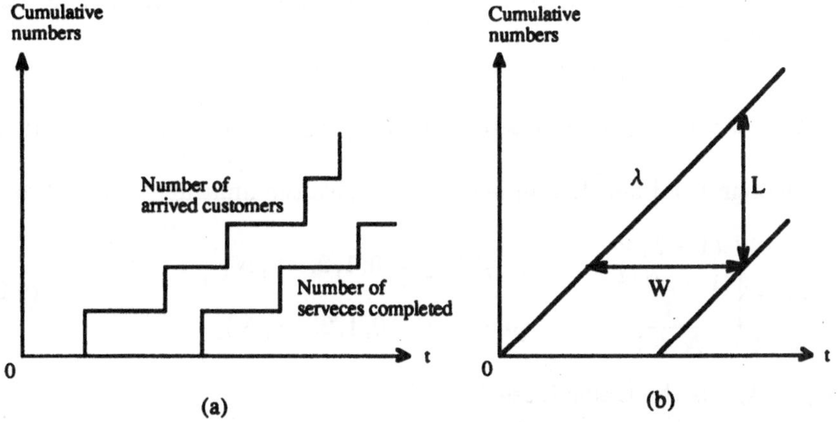

Fig. 9.2.1 Interpretation of Little's Formula $L = \lambda W$.

We recall Little's formulas in Eqs.(9.2.25) and (9.2.26). We present Fig. 9.2.1 with the cumulative numbers of customers who have arrived and customers completing service in (a) and the same figure for large T with emphasis in (b). Noting that both curves approach lines with slope λ for large T, we find that the cumulative numbers of customers who have arrived and customers completing service are λT and $\lambda(T-W)$ at time t, respectively, where W is the mean time spent in the system. We have the mean number of customers in the system after subtracting $\lambda(T-W)$ from λT:

$$L = \lambda T - \lambda(T-W) = \lambda W, \tag{9.2.27}$$

which is Little's Formula. We can also interpret Little's Formula $L_q = \lambda W_q$ similarly. For this interpretation, the intuitive proof is valid for not only M/M queueing models but also for generalized queueing models such as M/G/1. We should take note of Little's Formulas for varying λ. That is, if the arrival rate λ_k depends on state k, we should apply the *actual arrival rate* λ_a:

$$\lambda_a = \sum_{k=0}^{\infty} \lambda_k p_k \tag{9.2.28}$$

where p_k is the limiting probability. For instance, if we consider the queueing models with finite population, we should apply Eq.(9.2.28), since the arrival rate λ_k depends on state k (cf. Section 9.4).

9.2.2 M/M/1/N Queueing Models

In the preceding subsection we have assumed that the maximum queueing system capacity is infinite. However, we assume here that the maximum queueing system capacity is a finite N. That is, there are maximum $(N-1)$ customers waiting for service plus a customer being served.

For an M/M/1/N queueing model, all parameters are positive:

$$\lambda_k = \lambda > 0, \quad \mu_{k+1} = \mu > 0 \qquad (k = 0, 1, 2, \cdots, N-1). \tag{9.2.29}$$

From Theorem 6.4.3 and Example 6.4.4, we have the limiting probabilities:

$$p_j = \begin{cases} \dfrac{(1-\rho)\rho^j}{1-\rho^{N+1}} & (\rho \neq 1;\ j = 0, 1, 2, \cdots, N) \\ \dfrac{1}{N+1} & (\rho = 1;\ j = 0, 1, 2, \cdots, N), \end{cases} \tag{9.2.30}$$

where $\rho = \lambda/\mu$ is the traffic intensity.

The means of the number of customers in the system and in queue are given by

$$L = \sum_{j=0}^{N} j p_j = \rho \frac{1-(N+1)\rho^N + N\rho^{N+1}}{(1-\rho)(1-\rho^{N+1})}, \tag{9.2.31}$$

$$L_q = \sum_{j=1}^{N} (j-1)p_j = L - (1-p_0), \tag{9.2.32}$$

respectively, for $\rho \neq 1$. For $\rho = 1$, we have

$$L = \sum_{j=0}^{N} jp_j = \frac{N}{2}, \tag{9.2.33}$$

$$L_q = \sum_{j=1}^{N} (j-1)p_j = L - (1-p_0) = \frac{N(N-1)}{2(N+1)}. \tag{9.2.34}$$

Let U and V be the waiting time in queue and the service time, respectively, where $U+V$ is the time spent in the system. The conditional probability that there are j customers in the system given that a customer can join the system is given by

$$q_j = \frac{p_j}{1-p_N} \qquad (j = 0, 1, 2, \cdots, N-1), \tag{9.2.35}$$

since we cannot join the system if there are N customers in the system. The probability that a customer joined the queue before the $(n+1)$st customer has completed being serviced up until time t is given by

$$1 - \int_0^t \frac{\mu(\mu x)^n e^{-\mu x}}{n!} dx = \sum_{k=0}^{n} \frac{(\mu t)^k}{k!} e^{-\mu t} \qquad (n = 0, 1, \cdots, N-1), \tag{9.2.36}$$

which is the survival probability of the gamma distribution $GAM(\mu, n+1)$. The distribution of $U+V$ is given by using Eqs.(9.2.35) and (9.2.36):

$$P\{U + V \leq t\} = 1 - \sum_{j=0}^{N-1} q_j \sum_{k=0}^{j} \frac{(\mu t)^k}{k!} e^{-\mu t}. \tag{9.2.37}$$

Just as in Eq.(9.2.37), we have the distribution of U:

$$P\{U \leq t\} = 1 - \sum_{j=0}^{N-2} q_{j+1} \sum_{k=0}^{j} \frac{(\mu t)^k}{k!} e^{-\mu t}. \tag{9.2.38}$$

The mean time spent in the system is given by using Eq.(9.2.37):

$$W = E[U+V] = \sum_{j=0}^{N-1} q_j \frac{j+1}{\mu}$$

$$= \begin{cases} \dfrac{1-(N+1)\rho^N + N\rho^{N+1}}{\mu(1-\rho)(1-\rho^N)} & (\rho \neq 1) \\ \dfrac{N+1}{2\mu} & (\rho = 1). \end{cases} \tag{9.2.39}$$

Just as in Eq.(9.2.39), we have the mean waiting time:

$$W_q = E[U] = \sum_{j=0}^{N-2} q_{j+1} \frac{j+1}{\mu}$$

$$= \begin{cases} \rho \dfrac{1 - N\rho^{N-1} + (N-1)\rho^N}{\mu(1-\rho)(1-\rho^N)} & (\rho \neq 1) \\ \dfrac{N-1}{2\mu} & (\rho = 1). \end{cases} \tag{9.2.40}$$

Let us verify Little's Formulas for an M/M/1/N queueing model. Noting that it is impossible to join the queue when the queue is full (i.e., state N), we have the following *actual arrival rate* (cf. Eq.(9.2.28)):

$$\lambda_a = \sum_{j=0}^{N-1} p_j \lambda = (1 - p_N)\lambda. \tag{9.2.41}$$

Using the actual arrival rate λ_a, we have the following Little's Formulas:

$$L = \lambda_a W, \tag{9.2.42}$$

$$L_q = \lambda_a W_q, \tag{9.2.43}$$

(see Problem 9.4).

9.3 Multiple Server Queueing Models

Consider a multiple server queueing model in which there is a single queue and multiple servers (see Fig. 9.1.1). If the customers in queue find an empty server, the head customer in queue fills the empty service channel, where first come, first served rules are applied. In this section we discuss the following three multiple server queueing models:

- M/M/c/∞ Queueing Models
- M/M/c/c Queuing Models
- M/M/∞/∞ Queuing Models

9.3.1 M/M/c/∞ Queueing Models

Consider again a queue with infinite population. The arrival rate is $\lambda_k = \lambda$ for any integer $k \geq 0$. If there are k ($k \leq c$) customers being served simultaneously, the service rate is proportional to the number of customers being served, i.e., $\mu_k = k\mu$. However, if there are k ($k > c$) customers in the system, the service rate $\mu_k = c\mu$ since there are only c customers being served simultaneously and the remaining $(k - c)$ customers are in queue. Of course, μ is the service rate per single customer. Summarizing all the parameters, we assume the following for an M/M/c/∞ queueing model:

$$\lambda_k = \lambda \qquad (k = 0, 1, 2, \cdots), \tag{9.3.1}$$

$$\mu_k = \begin{cases} k\mu & (k = 0, 1, 2, \cdots, c) \\ c\mu & (k = c, c+1, c+2, \cdots). \end{cases} \tag{9.3.2}$$

Referring to Theorem 6.4.2, we have the necessary and sufficient condition for the limiting probabilities to exist:

$$\begin{aligned} \sum_{j=0}^{\infty} \prod_{k=1}^{j} \frac{\lambda_{k-1}}{\mu_k} &= \sum_{j=0}^{c-1} \frac{1}{j!}\left(\frac{\lambda}{\mu}\right)^j + \frac{1}{c!}\left(\frac{\lambda}{\mu}\right)^c \left[1 + \left(\frac{\lambda}{c\mu}\right) + \left(\frac{\lambda}{c\mu}\right)^2 + \cdots\right] \\ &= \sum_{j=0}^{c-1} \frac{u^j}{j!} + \frac{u^c}{c!} \sum_{n=0}^{\infty} \rho^n \\ &= \sum_{j=0}^{c-1} \frac{u^j}{j!} + \frac{u^c}{c!(1-\rho)} < \infty, \end{aligned} \tag{9.3.3}$$

which implies $\rho < 1$, where $u = \lambda/\mu$ and $\rho = u/c = \lambda/(c\mu)$. Assuming $\rho < 1$, we have the limiting probabilities:

$$p_0 = \left[\sum_{j=0}^{c-1} \frac{u^j}{j!} + \frac{u^c}{c!(1-\rho)}\right]^{-1}, \tag{9.3.4}$$

$$p_j = \begin{cases} \dfrac{u^j}{j!} p_0 & (j = 1, 2, \cdots, c) \\ \dfrac{u^c}{c!}\left(\dfrac{u}{c}\right)^{j-c} p_0 & (j = c, c+1, \cdots). \end{cases} \tag{9.3.5}$$

Let us calculate L_q, the mean number of customers in queue. Noting that the number of service channels is c, we have

$$L_q = \sum_{j=c}^{\infty} (j - c) p_j = p_0 \frac{u^c}{c!} (0 + 1\rho + 2\rho^2 + 3\rho^3 + \cdots) = \frac{p_0 u^c \rho}{c!(1-\rho)^2}. \tag{9.3.6}$$

By applying Little's Formulas, we have the mean waiting time:

$$W_q = \frac{L_q}{\lambda}. \tag{9.3.7}$$

Noting that the mean service time is $1/\mu$, we have the mean time spent in the system:

$$W = W_q + \frac{1}{\mu}. \tag{9.3.8}$$

Again applying Little's Formulas, we have $L = \lambda W$, the mean number of customers in the system.

The probability that a customer who just arrived has to wait for service is given by

$$P\{\text{a customer who arrived waits for service}\} = \sum_{j=c}^{\infty} p_j = \frac{u^c}{c!(1-\rho)} p_0, \tag{9.3.9}$$

which is called *Erlang's C Formula*:

$$C(c,u) = \frac{u^c}{c!(1-\rho)} p_0 = \frac{u^c/c!}{(1-\rho)\left[\sum_{j=0}^{c-1} \frac{u^j}{j!} + \frac{u^c}{c!(1-\rho)}\right]}. \tag{9.3.10}$$

Using Erlang's C Formula, we have the distribution of the waiting time:

$$P\{U \le t\} = 1 - C(c,u) e^{-c\mu t(1-\rho)}. \tag{9.3.11}$$

However, we omit the derivation of Eq.(9.3.11) since it is too complicated. The mean waiting time is given by

$$W_q = \frac{C(c,u)/\mu}{c(1-\rho)}, \tag{9.3.12}$$

which coincides with Eq.(9.3.7).

9.3.2 M/M/c/c Queueing Models

Consider an M/M/c/c queueing model in which the maximum queueing system capacity is equal to the number of service channels. That is, if a customer arrives at the system and he finds at least one empty server, he can join the system for service immediately. Otherwise, if he does not find any empty channels, he has to turn away without service. Such queueing models have been discussed in traffic theory, where the customers are the calls and the service channels are the (trunk) lines in traffic theory. If the lines are busy, the calls have to be canceled without service. Such a queueing model is called the *M/M/c Loss System* in traffic theory since we are interested in the (loss) probability that a call has to be canceled.

Referring to Theorem 6.4.3, we have the limiting probabilities irrespective of $\rho = u/c = \lambda/(c\mu)$:

$$p_j = \frac{u^j}{j!} p_0 \qquad (j = 1, 2, \cdots, c), \tag{9.3.13}$$

where

$$p_0 = \left[1 + \frac{u}{1!} + \frac{u^2}{2!} + \cdots + \frac{u^c}{c!}\right]^{-1}. \tag{9.3.14}$$

Rewriting Eq.(9.3.13), we have

$$p_j = \frac{u^j}{j!}e^{-u} \Big/ \left(\sum_{j=0}^{c} \frac{u^j}{j!}e^{-u}\right) \qquad (j = 0, 1, 2, \cdots, c), \tag{9.3.15}$$

which can be easily calculated by the Poisson distribution tables. In particular,

$$p_c = \frac{u^c/c!}{1 + u + u^2/2! + \cdots + u^c/c!} = B(c, u), \tag{9.3.16}$$

which is called *Erlang's B Formula* or *Erlang's Loss Formula* and is the probability that a customer has to turn away without service.

The actual arrival rate is given by

$$\lambda_a = \lambda(1 - p_c) = \lambda[1 - B(c, u)]. \tag{9.3.17}$$

Noting that $L_q = W_q = 0$ since there is no real queue in an M/M/c/c queueing model, we have the mean number of customers in the system:

$$L = \sum_{j=0}^{c} jp_j = up_0 \sum_{j=0}^{c-1} \frac{u^j}{j!} = u[1 - B(c, u)]. \tag{9.3.18}$$

We can easily verify Little's Formulas:

$$W = \frac{L}{\lambda_a} = \frac{1}{\mu}. \tag{9.3.19}$$

We have obtained interesting results for an M/M/c/c queueing model. However, all the results in this subsection are also valid for an M/G/c/c queueing model, where we interpret that $1/\mu$ is the mean service time for an M/G/c/c queueing model. Such a queueing system is called a *robust* system.

9.3.3 M/M/∞/∞ Queueing Models

For an M/M/∞/∞ queueing model, all the arriving customers can be served immediately because the number of service channels is infinite. We can hardly observe such queueing models in practice. However, we can approximate such queueing models by regarding as c is large enough for an M/M/c/c queuing model. In particular, the analytical results are quite simple and easy to calculate in practice. Referring to Example 6.4.3, we have the limiting probabilities:

$$p_j = \frac{u^j}{j!}e^{-u} \quad (j = 0, 1, 2, \cdots) \tag{9.3.20}$$

which is the Poisson distribution $X \sim POI(u)$.

Using the fact that $X \sim POI(u)$, we have the mean number of customers in the system:

$$L = u. \tag{9.3.21}$$

Applying Little's Formula, we have

$$W = u/\lambda = 1/\mu, \tag{9.3.22}$$

which is naturally the mean service time. Noting that there are no queues, we have $L_q = W_q = 0$.

As shown in Example 3.4.1, we have derived the analytical results for an M/G/∞/∞ queueing model. That is, the probability that there are j customers being served at time t given that there are no customers at $t = 0$ is given by

$$P\{j \text{ customers being served at time } t\} = \frac{(p\lambda t)^j}{j!}e^{-p\lambda t} \quad (j = 0, 1, 2, \cdots), \tag{9.3.23}$$

i.e., the probability that j customers being served at time t follows a nonhomogeneous Poisson process with the mean value function

$$p\lambda t = \lambda \int_0^t [1 - G(x)]dx, \tag{9.3.24}$$

where $G(t)$ is an arbitrary time distribution. If $t \rightarrow \infty$ in Eq.(9.3.23), we have

$$P\{j \text{ customers being served in the steady state}\} = \frac{u^j}{j!}e^{-u} \quad (j = 0, 1, 2, \cdots), \tag{9.3.25}$$

where

$$\lim_{t\rightarrow\infty} p\lambda t = \lambda \int_0^\infty [1 - G(x)]dx = \frac{\lambda}{\mu} = u. \tag{9.3.26}$$

The probabilities in Eq.(9.3.25) have been given by p_j in Eq.(9.3.20).

In particular, if we assume the exponential service time distribution $G(t) = 1 - e^{-\mu t}$, we have

$$p\lambda t = \lambda \int_0^t [1 - G(x)]dx = \frac{\lambda}{\mu}(1 - e^{-\mu t}), \tag{9.3.27}$$

which is the mean value function of the nonhomogeneous Poisson process. Of course, the results in Eq.(9.3.23), having the mean value function in Eq.(9.3.27), can be analytically solved by Kolmogorov's forward equations in Eqs.(6.4.5) and (6.4.7), where $\lambda_k = \lambda$, $\mu_{k+1} = (k+1)\mu$, $P_{00}(0) = 1$, and $P_{0j}(0) = 0$ $(j \geq 1)$.

9.4 Queues with a Finite Population

So far we have discussed M/M queueing models with an infinite population. That is, the arrival rate λ_k is constant irrespective of state k. In this section we focus on M/M queueing models with a finite population. Typical examples of such queueing models occur in *(machine) repairman problems.* Throughout this section we use the terms from repairman problems. That is, we have finite K *machines* which can operate. Once a machine breaks down, the machine is repaired immediately or waits to be repaired if the *repairmen* are busy. We assume that the *failure* (arrival) *rate* is λ for each machine and the *repair* (service) *rate* is μ for each repairman. If there are n machines operating, where $n = 1, 2, \cdots, K$, then the probability that one of n machines breaks down for a small interval $h > 0$ is given by

$$\binom{n}{1}(1 - e^{-\lambda h})e^{-(n-1)\lambda h} = n\lambda h + o(h). \tag{9.4.1}$$

That is, the failure rate varies depending on the number of machines operating.

In this section we discuss the following queueing models with a finite population:

- M/M/1/K/K Queueing Models
- M/M/c/K/K Queueing Models
- M/M/c/c/c Queueing Models

9.4.1 M/M/1/K/K Queueing Models

Let state j be the number of failed machines, where $j = 1, 2, \cdots, K$. Noting that the failure rate is λ for each machine and the repair rate is μ for each repairman, we have

$$\lambda_j = (K - j)\lambda \qquad (j = 0, 1, 2, \cdots, K - 1), \tag{9.4.2}$$

$$\mu_j = \mu \qquad (j = 1, 2, \cdots, K). \tag{9.4.3}$$

Referring to Theorem 6.4.3, we have

$$p_j = \frac{K!}{(K - j)!}\left(\frac{\lambda}{\mu}\right)^j p_0 \qquad (j = 1, 2, \cdots, K), \tag{9.4.4}$$

where

$$p_0 = \left[\sum_{j=0}^{K} \frac{K!}{(K - j)!}\left(\frac{\lambda}{\mu}\right)^j\right]^{-1}. \tag{9.4.5}$$

The probability that there is at least a machine being repaired is given by

$$\begin{aligned}
&P\{\text{at least a machine being repaired}\}\\
&= \sum_{j=1}^{K} p_j\\
&= 1 - p_0\\
&= 1 - \frac{1}{\sum_{j=0}^{K} \frac{K!}{(K-j)!}\left(\frac{\lambda}{\mu}\right)^j}\\
&= 1 - B(K, \mu/\lambda),
\end{aligned} \tag{9.4.6}$$

where $B(K,\mu/\lambda)$ is *Erlang's B Formula* in Eq.(9.3.16).

The actual failure (arrival) rate is given by

$$\lambda_a = \sum_{j=0}^{K-1} \lambda_j p_j = (1-p_0)\mu \tag{9.4.7}$$

(see Problem 9.11). We can also obtain the actual failure rate as follows: A machine fails, waits for repair, and is repaired, which is a cycle, where the respective means are $1/\lambda$, W_q, and $1/\mu$. Noting that there are K machines, we have

$$\lambda_a = \frac{K}{1/\lambda + W_q + 1/\mu}. \tag{9.4.8}$$

From Eqs.(9.4.7) and (9.4.8) with respect to W_q, we have

$$W_q = \frac{K}{(1-p_0)\mu} - \frac{1}{\lambda} - \frac{1}{\mu}. \tag{9.4.9}$$

The mean time spent in the system is given by

$$W = W_q + \frac{1}{\mu} = \frac{K}{(1-p_0)\mu} - \frac{1}{\lambda}. \tag{9.4.10}$$

Using λ_a in Eq.(9.4.7) and Little's Formulas, we have

$$L = \lambda_a W\ , \tag{9.4.11}$$

$$L_q = \lambda_a W_q\ . \tag{9.4.12}$$

Of course, we can calculate L and L_q as follows:

$$L = \sum_{j=0}^{K} j p_j\ , \tag{9.4.13}$$

$$L_q = \sum_{j=1}^{K} (j-1) p_j\ , \tag{9.4.14}$$

(see Problem 9.11).

9.4.2 M/M/c/K/K Queueing Models

Consider a similar queueing model with c repairmen. We again define state j such that the number of failed machines is j, where $j = 0, 1, 2, \cdots, K$. All the parameters are given by

$$\lambda_j = (K - j)\lambda \qquad (j = 0, 1, 2, \cdots, K - 1), \tag{9.4.15}$$

$$\mu_j = \begin{cases} j\mu & (j = 1, 2, \cdots, c) \quad (9.4.16) \\ c\mu & (j = c, c + 1, \cdots, K). \quad (9.4.17) \end{cases}$$

Referring to Theorem 6.4.3, we have

$$p_j = \begin{cases} \binom{K}{j}\left(\frac{\lambda}{\mu}\right)^j p_0 & (j = 0, 1, 2, \cdots, c) \\ \frac{j!}{c!c^{j-c}}\binom{K}{j}\left(\frac{\lambda}{\mu}\right)^j p_0 & (j = c, c + 1, \cdots, K), \end{cases} \tag{9.4.18}$$

where

$$p_0 = \left[\sum_{j=0}^{c}\binom{K}{j}\left(\frac{\lambda}{\mu}\right)^j + \sum_{j=c+1}^{K}\frac{j!}{c!c^{j-c}}\binom{K}{j}\left(\frac{\lambda}{\mu}\right)^j\right]^{-1}. \tag{9.4.19}$$

The mean numbers of customers in the system and in queue are given by

$$L = \sum_{j=0}^{K} jp_j\ , \tag{9.4.20}$$

$$L_q = \sum_{j=c}^{K}(j - c)p_j\ , \tag{9.4.21}$$

respectively. However, we can derive the mean waiting time W_q as we have shown in Eq.(9.4.9) as follows: The actual failure rate is given by

$$\lambda_a = \frac{K}{1/\lambda + W_q + 1/\mu}. \tag{9.4.22}$$

Using Little's Formulas, we have the mean waiting time:

$$W_q = \frac{L_q}{\lambda_a} = \left(\frac{1}{\lambda} + W_q + \frac{1}{\mu}\right)\left(\frac{L_q}{K}\right), \tag{9.4.23}$$

which implies

$$W_q = \frac{L_q\left(\frac{1}{\lambda} + \frac{1}{\mu}\right)}{K - L_q}. \tag{9.4.24}$$

That is, W_q can be derived from L_q, and vice versa. The mean time spent in the system is given by

$$W = W_q + \frac{1}{\mu}, \tag{9.4.25}$$

which implies

$$L = \lambda_a W, \tag{9.4.26}$$

from Little's Formulas.

Using Eqs.(9.4.18) and (9.4.19), we introduce the following quantities which specify a repairman problem:

- Mean number of machines being repaired

$$L_r = \sum_{j=0}^{c} j p_j + c \sum_{j=c+1}^{K} p_j \ , \tag{9.4.27}$$

- Mean number of repairmen idle

$$L_i = \sum_{j=0}^{c} (c - j) p_j \ , \tag{9.4.28}$$

- Coefficient of loss for machines

$$M_q = \frac{\text{[mean number of machines in queue]}}{\text{[number of machines]}} = \frac{L_q}{K}, \tag{9.4.29}$$

- Coefficient of repair for machines

$$M_r = \frac{\text{[mean number of machines being repaired]}}{\text{[number of machines]}} = \frac{L_r}{K}, \tag{9.4.30}$$

- Coefficient of loss for repairmen

$$R_i = \frac{\text{[mean number of repairmen idle]}}{\text{[number of repairmen]}} = \frac{L_i}{c}, \tag{9.4.31}$$

- Coefficient of operation for machines

$$M_w = \frac{\text{[mean number of machines operating]}}{\text{[number of machines]}} = 1 - \frac{L}{K}, \tag{9.4.32}$$

where

$$M_q + M_r + M_w = 1, \tag{9.4.33}$$

(see Problem 9.12). Such measures in Eqs.(9.4.27)-(9.4.32) are appropriate for evaluating a repairman problem in practice.

9.4.3 M/M/c/c/c Queueing Models

Applying the results in the preceding subsection by assuming $K = c$, we have

$$p_j = \binom{c}{j}\left(\frac{\lambda}{\lambda+\mu}\right)^j\left(\frac{\mu}{\lambda+\mu}\right)^{c-j} \qquad (j = 0, 1, 2, \cdots, c), \tag{9.4.34}$$

which is the binomial distribution $X \sim B\left(c, \frac{\lambda}{\lambda+\mu}\right)$. The mean number of customers in the system is given by

$$L = \frac{c\lambda}{\lambda+\mu}. \tag{9.4.35}$$

The actual failure rate is given by

$$\lambda_a = \sum_{j=0}^{c-1} p_j\lambda_j = \frac{c\lambda\mu}{\lambda+\mu}, \tag{9.4.36}$$

and the mean time spent in the system is given by

$$W = \frac{1}{\mu} = \frac{L}{\lambda_a}, \tag{9.4.37}$$

which verifies Little's Formulas.

We have just derived the limiting probabilities and other quantities for an M/M/c/c/c queueing model. However, if we restrict ourselves to this model, we can understand that each machine is maintained by each repairman and there are c machines in parallel. We can obtain the analytical results, i.e., the transient solutions for an M/M/c/c/c queueing model in general as follows: Referring to Example 6.4.5, we have

$$P_{00}(t) = \frac{\mu}{\lambda+\mu} + \frac{\lambda}{\lambda+\mu}e^{-(\lambda+\mu)t}, \tag{9.4.38}$$

$$P_{01}(t) = \frac{\lambda}{\lambda+\mu} - \frac{\lambda}{\lambda+\mu}e^{-(\lambda+\mu)t}, \tag{9.4.39}$$

which are the probabilities that a machine is operating and being repaired at time t, respectively, given that it is operating at $t = 0$. Since there are c machines in parallel, and assuming that all c machines are operating at $t = 0$, we have the following probabilities that j machines are being repaired at time t:

$$Q_{0j}(t) = \binom{c}{j}[P_{00}(t)]^{c-j}[P_{01}(t)]^j \qquad (j = 0, 1, 2, \cdots, c), \tag{9.4.40}$$

which are the transient solutions for an M/M/c/c/c queueing model. Of course, if $t \to \infty$, we have

$$p_j = \lim_{t\to\infty} Q_{0j}(t) \qquad (j = 0, 1, 2, \cdots, c), \tag{9.4.41}$$

which have been given in Eq.(9.4.34).

In queueing theory, it is very difficult or impossible to obtain the transient solutions analytically. Fortunately, we have successfully obtained the transient solutions from a different viewpoint. We have also shown the transient solutions for an $M/G/\infty$ queueing model from the viewpoint of the Poisson process.

9.5 Problems 9

9.1 The take-out counter at a hamburger shop is served by one attendant, and can be formulated by an M/M/1/∞ queueing model. Customers arrive according to a Poisson Process with 45 persons per hour, and are served with the mean service time of 1 min. Determine (i) the probability that the attendant is busy, (ii) the mean number of customers in queue, (iii) the mean waiting time for service, and (iv) the probability that a customer has to wait more than 4 min. in queue.

9.2 A secretary receives her word processing work. This model can be formulated by an M/M/1/∞ queueing model. The jobs arrive with the Poisson rate of six jobs per hour, and are served by the mean service time of 8 min. per job. Determine (i) the probability that the secretary is busy, (ii) the mean number of jobs in the system, (iii) the mean time spent in the system, and (iv) the probability that an arriving job will be completed in under 40 min.

9.3 Verify that, for an M/M/1/∞ queueing model, the probability that there are n or more customers in the system is ρ^n.

9.4 Verify Little's Formulas $L = \lambda_a W$ and $L_q = \lambda_a W_q$ for an M/M/1/N queueing model when $\rho \neq 1$ by using the results in Eqs.(9.2.33), (9.2.34), (9.2.39), (9.2.40), and (9.2.41).

9.5 A dental clinic has a waiting room for 4 persons and is served by a dentist. This model can be formulated by an M/M/1/5 queueing model. The patients arrive at the Poisson rate of 2 persons per hour and the dentist spends a mean of 20 min. per patient. Determine (i) the probability that the dentist is busy, (ii) the mean number of the patients in the system, and (iii) the mean waiting time.

9.6 For an M/M/c/N queueing model ($c \leq N$), we assume that the arrival rate is λ and the service rate is μ. Determine (i) the limiting probabilities p_j for $\rho = \lambda/\mu = 1$ or $\rho \neq 1$, (ii) the mean number of customers in queue L_q, (iii) the actual arrival rate, and (iv) W_q, W, and L from Little's Formulas $L_q = \lambda_a W_q$, $L = \lambda_a W$, and the formula $W = W_q + 1/\mu$.

9.7 (*Continuation*) Verify that the results for an M/M/1/N queueing model are those of a special M/M/c/N queueing model with $c = 1$.

9.8 Consider a telephone system with 4 lines, which can be formulated by an M/M/4/4 queueing model. Determine (i) the probability that all the lines are busy, i.e., p_4 for varying $u = \lambda/\mu = 1.0, 1.1, \cdots, 2.0$, and (ii) how we decide the utilization factor u if the loss probability is less than 5%.

9.9 For an M/M/1/6/6 queueing model with finite population, calculate the following quantities by assuming $\lambda/\mu = 0.2$: (i) M_w, (ii) M_q, (iii) M_r, and (iv) R_i.

9.10 (*Continuation*) For an M/M/2/12/12 queueing model with finite population, calculate the quantities (i), (ii), (iii), and (iv) in Problem 9.9 and compare them to the measures in Problem 9.9.

9.11 Consider an M/M/1/K/K queueing model with finite population. Derive the following:

(i) The actual arrival rate is given by

$$\lambda_a = \sum_{j=0}^{K-1} \lambda_j p_j = \mu(1 - p_0).$$

(ii) The mean number of customers in the system is given by

$$L = \sum_{j=1}^{K} j p_j = K - \frac{\mu}{\lambda}(1 - p_0).$$

(iii) The mean time spent in the system is given by

$$W = \frac{L}{\lambda_a} = \frac{K}{\mu(1 - p_0)} - \frac{1}{\lambda}$$

from Little's Formulas.

9.12 For an M/M/c/K/K queueing model with finite population, verify that

$$M_q + M_r + M_w = 1.$$

Appendix A

Laplace-Stieltjes Transforms

A.1 Laplace-Stieltjes Transforms

Let $F(t)$ be a well-defined function of t specified for $t \geq 0$ and s be a complex number. If the following Stieltjes integral:

$$F^*(s) = \int_0^\infty e^{-st} dF(t) \tag{A.1.1}$$

converges on some s_0, the Stieltjes integral (A.1.1) converges on s such that $\Re(s) > \Re(s_0)$. The integral (A.1.1) is called the *Laplace-Stieltjes transform* of $F(t)$. If the real function $F(t)$ can be expressed in terms of the following integral:

$$F(t) = \int_0^t dF(x) = \int_0^t f(x)dx, \tag{A.1.2}$$

then

$$F^*(s) = \int_0^\infty e^{-st} dF(t) = \int_0^\infty e^{-st} f(t)dt, \tag{A.1.3}$$

which is called the *Laplace transform* of $f(t)$.

Noting that $F(t)$ is in one-to-one correspondence with $F^*(s)$ (ref. Theorem 2.2.2 and Table 2.2.2), $F(t)$ can be uniquely specified by $F^*(s)$. The *inversion formula* for obtaining $F(t)$ from $F^*(s)$ can be given by

$$F(t) = \lim_{c \to \infty} \frac{1}{2\pi i} \int_{b-ic}^{b+ic} \frac{e^{st}}{s} F^*(s) ds, \tag{A.1.4}$$

where $i = \sqrt{-1}$ is an imaginary unit, $b > \max(\sigma, 0)$ and σ is a radius of convergence.

The following two theorems are well-known and of great use as the limit theorems for the Laplace-Stieltjes transform $F^*(s)$ of $F(t)$.

Theorem A.1 (*An Abelian Theorem*) If for some non-negative number α,

$$\lim_{t\to\infty} \frac{F(t)}{t^\alpha} = \frac{C}{\Gamma(\alpha+1)}, \tag{A.1.5}$$

then

$$\lim_{s\to+0} s^\alpha F^*(s) = C, \tag{A.1.6}$$

where $\Gamma(k) = \int_0^\infty e^{-x} x^{k-1} dx$ is a gamma function of order k defined in Eq.(2.4.13).

Theorem A.2 (*A Tauberian Theorem*) If $F(t)$ is non-decreasing and the Laplace-Stieltjes transform

$$F^*(s) = \int_0^\infty e^{-st} dF(t) \tag{A.1.7}$$

converges for $\Re(s) > 0$, and if for some non-negative number α,

$$\lim_{s\to+0} s^\alpha F^*(s) = C, \tag{A.1.8}$$

then

$$\lim_{t\to\infty} \frac{F(t)}{t^\alpha} = \frac{C}{\Gamma(\alpha+1)}. \tag{A.1.9}$$

Exampe A.1.1 (*Elementery Renewal Theorem*) The Laplace-Stieltjes transform of the renewal function is given in Eq.(4.2.15):

$$M^*(s) = \frac{F^*(s)}{1-F^*(s)},$$

where $F^*(s)$ is the Laplace-Stieltjes transform of $F(t)$. Let us apply Theorem A.2.

We have

$$\lim_{s\to+0} sM^*(s) = \lim_{s\to+0} \frac{F^*(s)}{[1-F^*(s)]/s} = \frac{1}{\mu},$$

since

$$\lim_{s\to+0} \frac{1-F^*(s)}{s} = \lim_{s\to+0} \frac{\mu s + \frac{1}{2}(\sigma^2+\mu^2)s^2 + o(s^2)}{s} = \mu,$$

and

$$\lim_{s\to 0} F^*(s) = 1.$$

That is, from the Tauberian Theorem, we have

$$\lim_{t\to\infty} \frac{M(t)}{t} = \frac{1}{\mu}.$$

A.2 Properties of Laplace-Stieltjes Transforms

For the Laplace-Stieltjes transform, we have the following relationship:

$$F^*(s) = \int_0^\infty e^{-st} dF(t) = s \int_0^\infty e^{-st} F(t) dt. \tag{A.2.1}$$

That is, the Laplace-Stieltjes transform $F^*(s)$ can be obtained by s times the Laplace transform of $F(t)$. We can easily obtain the Laplace-Stieltjes transforms from the corresponding Laplace transforms, since most textbooks only discuss the latter.

Table A.1 shows the the general properties of the Laplace-Stieltjes transforms. The general properties in Table A.1 can be applied in practice. Besides the general properties in Table A.1, Table A.2 shows the important formulas for the Laplace-Stieltjes transforms. Such formulas in Table A.2 can be applied to derive the Laplace-Stieltjes transform $F^*(s)$ from $F(t)$, and vice versa.

Table A.1 General properties of the Laplace-Stieltjes transforms.

$F(t)$	$F^*(s) = \int_0^\infty e^{-st} dF(t)$
$F_1(t) + F_2(t)$	$F_1^*(s) + F_2^*(s)$
$aF(t)$	$aF^*(s)$
$F(t-a) \quad (a > 0)$	$e^{-sa} F^*(s)$
$F(at) \quad (a > 0)$	$F^*(s/a)$
$e^{-at} F(t) \quad (a > 0)$	$\frac{s}{s+a} F^*(s+a)$
$F'(t) = \frac{dF(t)}{dt}$	$s[F^*(s) - F(0)]$
$tF'(t)$	$-s \frac{dF^*(s)}{ds}$
$\int_0^t F(x) dx$	$\frac{1}{s} F^*(s)$
$\int_0^t \cdots \int_0^t F(t)(dt)^n$	$\frac{1}{s^n} F^*(s)$
$\lim_{t \to +0} F(t)$	$\lim_{s \to \infty} F^*(s)$
$\lim_{t \to \infty} F(t)$	$\lim_{s \to +0} F^*(s)$

Table A.2 Formulas of the Laplace-Stieltjes Transforms

$F(t)$	$F^*(s) = \int_0^\infty e^{-st}dF(t)$
$\delta(t-a)^\dagger \quad (a>0)$	se^{-sa}
$1(t-a)^\ddagger \quad (a>0)$	e^{-sa}
$1(t)$	1
t	$\frac{1}{s}$
$t^n \quad (n:\text{a positive integer})$	$\frac{n!}{s^n}$
$t^\alpha \quad (\alpha>-1)$	$\frac{\Gamma(\alpha+1)}{s^\alpha}$
$e^{-\alpha t} \quad (\alpha>0)$	$\frac{s}{s+\alpha}$
$te^{-\alpha t} \quad (\alpha>0)$	$\frac{s}{(s+\alpha)^2}$
$t^ne^{-\alpha t} \quad (\alpha>0)$	$\frac{n!s}{(s+\alpha)^{n+1}}$
$t^\beta e^{-\alpha t} \quad (\alpha>0,\beta>-1)$	$\frac{s\Gamma(\beta+1)}{(s+\alpha)^{\beta+1}}$
$\cos\alpha t$	$\frac{s^2}{s^2+\alpha^2} \quad (\Re(s)>\mid\alpha\mid)$
$\sin\alpha t$	$\frac{s\alpha}{s^2+\alpha^2} \quad (\Re(s)>\mid\alpha\mid)$
$\cosh\alpha t$	$\frac{s^2}{s^2-\alpha^2} \quad (\Re(s)>\mid\alpha\mid)$
$\sinh\alpha t$	$\frac{s\alpha}{s^2-\alpha^2} \quad (\Re(s)>\mid\alpha\mid)$
$\log t$	$-\gamma-\log s \quad \S$

† Dirac's delta function.
‡ Heaviside's unit function.
§ $\gamma = 0.57721\cdots$, Euler's constant.

A.3 Applications to Distributions

We discuss the Laplace-Stieltjes transforms of the distributions. As shown in Section 2.3, we have introduced six common discrete distributions. We derive the Laplace-Stieltjes transforms for a few discrete distributions.

Example A.3.1 (*Binomial distribution*)

$$F_X^*(s) = \sum_{x=0}^{n} e^{-sx} \binom{n}{x} p^x q^{n-x} = (pe^{-s} + q)^n \,. \tag{A.3.1}$$

Applying the formulas for the moments, we have

$$E[X] = (-1)\frac{dF_X^*(s)}{ds}\bigg|_{s=0} = np, \tag{A.3.2}$$

$$E[X^2] = (-1)^2\frac{d^2F_X^*(s)}{ds^2}\bigg|_{s=0} = n(n-1)p^2 + np, \tag{A.3.3}$$

which imply

$$Var(X) = E[X^2] - E[X]^2 = npq. \tag{A.3.4}$$

Example A.3.2 (*Geometric distribution*)

$$F_X^*(s) = \sum_{x=1}^{\infty} e^{-sx} pq^{x-1} = \frac{pe^{-s}}{1 - qe^{-s}} \,. \tag{A.3.5}$$

Example A.3.3 (*Negative binomial distribution*)

$$F_X^*(s) = \sum_{x=r}^{\infty} e^{-sx} \binom{x-1}{x-r} p^r q^{x-r} = \left[\frac{pe^{-s}}{1 - qe^{-s}}\right]^r \,. \tag{A.3.6}$$

As shown in Example 2.3.2, if $X_1, X_2, \cdots, X_r$ are independent and identically distributed random variables with $X_i \sim GEO(p)$, the random variable $S_r = X_1 + X_2 + \cdots + X_r$ is distributed with the negative binomial distribution $S_r \sim NB(p, r)$. Here we have verified this fact by using the Laplace-Stieltjes transforms.

We next show the Laplace-Stieltjes transforms for the continuous time distributions in Section 2.4.

Example A.3.4 (*Exponential distribution*)

$$F_X^*(s) = \int_0^\infty e^{-st} dF_X(t) = \int_0^\infty e^{-st} \lambda e^{-\lambda t} dt = \frac{\lambda}{s + \lambda}. \tag{A.3.7}$$

Applying the formulas for the moments, we have

$$E[X] = (-1)\frac{dF_X^*(s)}{ds}\bigg|_{s=0} = 1/\lambda, \tag{A.3.8}$$

$$E[X^2] = (-1)^2 \frac{d^2 F_X^*(s)}{ds^2}\bigg|_{s=0} = 2/\lambda^2, \tag{A.3.9}$$

which imply

$$Var(X) = \frac{2}{\lambda^2} - \frac{1}{\lambda^2} = \frac{1}{\lambda^2}. \tag{A.3.10}$$

Example A.3.5 (*Gamma distribution*)

$$F_X^*(s) = \int_0^\infty e^{-st} dF_X(t) = \int_0^\infty \frac{\lambda^k t^{k-1} e^{-(\lambda+s)t}}{\Gamma(k)} dt = \left(\frac{\lambda}{s+\lambda}\right)^k. \tag{A.3.11}$$

Applying the formulas for the moments, we have

$$E[X] = \frac{k}{\lambda}, \tag{A.3.12}$$

$$Var(X) = \frac{k}{\lambda^2}. \tag{A.3.13}$$

Example A.3.6 (*Equilibrium distribution*)

In Chapter 4, we have introduced the equilibrium distribution in Eq.(4.3.44):

$$F_e(t) = \frac{1}{\mu} \int_0^t [1 - F(y)]\ dy, \tag{A.3.14}$$

where μ is the mean of $F(t)$. The Laplace-Stieltjes transform of $F_e(t)$ is given by

$$\begin{aligned} F_e^*(s) &= \int_0^\infty e^{-st}\ dF_e(t) \\ &= \frac{1}{\mu s} \int_0^\infty e^{-st}\ d[1 - F(t)] \quad \text{(see Table A.1)} \\ &= \frac{1}{\mu s} [1 - F^*(s)]. \end{aligned} \tag{A.3.15}$$

A.4 Applications to Differential Equations

Example A.4.1 (*Example 6.4.5*) We have discussed a two-state Markov chain whose Kolmogorov's forward equations are given by

$$P_{00}'(t) = -\lambda P_{00}(t) + \mu P_{01}(t), \tag{A.4.1}$$

$$P_{01}'(t) = \lambda P_{00}(t) - \mu P_{01}(t), \tag{A.4.2}$$

with the initial conditions that $P_{00}(0) = 1$ and $P_{01}(0) = 0$. Let

$$P_{0j}^*(s) = \int_0^\infty e^{-st}\, dP_{0j}(t) \tag{A.4.3}$$

be the Laplace-Stieltjes transforms of $P_{0j}(t)$ $(j = 0, 1)$. Noting the initial conditions and using Table A.1, we have the Laplace-Stieltjes transform expressions for Eqs.(A.4.1) and (A.4.2):

$$sP_{00}^*(s) - s = -\lambda P_{00}^*(s) + \mu P_{01}^*(s), \tag{A.4.4}$$

$$sP_{01}^*(s) = \lambda P_{00}^*(s) - \mu P_{01}^*(s), \tag{A.4.5}$$

whose solutions are given by

$$P_{00}^*(s) = \frac{s+\mu}{s+\lambda+\mu} = \frac{s}{s+\lambda+\mu} + \frac{\mu}{\lambda+\mu} \cdot \frac{\lambda+\mu}{s+\lambda+\mu}, \tag{A.4.6}$$

$$P_{01}^*(s) = \frac{\lambda}{s+\lambda+\mu} = \frac{\lambda}{\lambda+\mu} \cdot \frac{\lambda+\mu}{s+\lambda+\mu}. \tag{A.4.7}$$

Applying the formulas in Table A.2, we have the following inversions of the Laplace-Stieltjes transforms:

$$P_{00}(t) = \frac{\mu}{\lambda+\mu} + \frac{\lambda}{\lambda+\mu} e^{-(\lambda+\mu)t}, \tag{A.4.8}$$

$$P_{01}(t) = \frac{\lambda}{\lambda+\mu} - \frac{\lambda}{\lambda+\mu} e^{-(\lambda+\mu)t}, \tag{A.4.9}$$

which have been given in Example 6.4.5.

Example A.4.2 (M/M/2/2/2 *queueing model*) We have discussed an M/M/c/c/c queueing model with finite population in Section 9.4 in general. We restrict ourselves to a case of $c = 2$ (i.e., 2 machines). Let $Q_{0j}(t)$ be the probabilities that $(2-j)$ machines are operating at time t given that 2 machines are operating at $t = 0$, where $j = 0, 1, 2$. Kolmogorov's forward equations are given by

$$Q_{00}'(t) = -2\lambda Q_{00}(t) + \mu Q_{01}(t), \tag{A.4.10}$$

$$Q_{01}'(t) = 2\lambda Q_{00}(t) - (\lambda+\mu) Q_{01}(t) + 2\mu Q_{02}(t), \tag{A.4.11}$$

$$Q_{02}'(t) = \lambda Q_{01}(t) - 2\mu Q_{02}(t), \tag{A.4.12}$$

with the initial conditions $Q_{00}(0) = 1$ and $Q_{0j}(0) = 0$ $(j = 1, 2)$. Let

$$Q_{0j}^*(s) = \int_0^\infty e^{-st}\, dQ_{0j}(t) \tag{A.4.13}$$

be the Laplace-Stieltjes transforms of $Q_{0j}(t)$ $(j = 0, 1, 2)$. The Laplace-Stieltjes transform expressions for Eqs.(A.4.10), (A.4.11), and (A.4.12) are given by

$$sQ_{00}^*(s) - s = -2\lambda Q_{00}^*(s) + \mu Q_{01}^*(s), \tag{A.4.14}$$

$$sQ_{01}^*(s) = 2\lambda Q_{00}^*(s) - (\lambda+\mu)Q_{01}^*(s) + 2\mu Q_{02}^*(s), \tag{A.4.15}$$

$$sQ_{02}^*(s) = \lambda Q_{01}^*(s) - 2\mu Q_{02}^*(s). \tag{A.4.16}$$

Solving with respect to $Q_{00}^*(s)$, we have

$$\begin{aligned} Q_{00}^*(s) &= \frac{s^2 + (\lambda+3\mu)s + 2\mu^2}{(s+\lambda+\mu)\,[s+2(\lambda+\mu)]} \\ &= 1 - \frac{2\lambda\mu}{(\lambda+\mu)^2}\cdot\frac{\lambda+\mu}{s+\lambda+\mu} - \frac{\lambda^2}{(\lambda+\mu)^2}\cdot\frac{2(\lambda+\mu)}{s+2(\lambda+\mu)}. \end{aligned} \tag{A.4.17}$$

where the last equation has been derived by applying the *partial fraction expansion* (or *decomposition*). Applying the formulas in Table A.2, we have the following inversion of the Laplace-Stieltjes transform in Eq.(A.4.17):

$$\begin{aligned} Q_{00}(t) &= 1 - \frac{2\lambda\mu}{(\lambda+\mu)^2}\left[1 - e^{-(\lambda+\mu)t}\right] - \frac{\lambda^2}{(\lambda+\mu)^2}\left[1 - e^{-2(\lambda+\mu)t}\right] \\ &= \left(\frac{\mu}{\lambda+\mu}\right)^2 + \frac{2\lambda\mu}{(\lambda+\mu)^2}e^{-(\lambda+\mu)t} + \left(\frac{\lambda}{\lambda+\mu}\right)^2 e^{-2(\lambda+\mu)t} \\ &= \left[\frac{\mu}{\lambda+\mu} + \frac{\lambda}{\lambda+\mu}e^{-(\lambda+\mu)t}\right]^2 \\ &= [P_{00}(t)]^2, \end{aligned} \tag{A.4.18}$$

where $P_{00}(t)$ has been given in Eq.(A.4.8). Of course, Eq.(A.4.18) can be easily derived by considering two M/M/1/1/1 queueing models in parallel (cf. Section 9.4). Similarly, we have the following Laplace-Stieltjes transforms:

$$Q_{01}^*(s) = \frac{2\lambda(s+2\mu)}{(s+\lambda+\mu)\,[s+2(\lambda+\mu)]}, \tag{A.4.19}$$

$$Q_{02}^*(s) = \frac{2\lambda^2}{(s+\lambda+\mu)\,[s+2(\lambda+\mu)]}, \tag{A.4.20}$$

which imply

$$Q_{01}(t) = 2P_{00}(t)P_{01}(t), \tag{A.4.21}$$

$$Q_{02}(t) = [P_{01}(t)]^2, \tag{A.4.22}$$

where $P_{00}(t)$ and $P_{01}(t)$ have been given in Eqs.(A.4.8) and (A.4.9), respectively.

A.5 Applications to Renewal Functions

In Chapter 4 we have developed renewal processes. Let $\{N(t),\ t \geq 0\}$ be a renewal process with interarrival distribution $F(t)$. The renewal function is given by

$$M(t) = E\left[N(t)\right] = \sum_{n=1}^{\infty} F^{(n)}(t). \tag{A.5.1}$$

The Laplace-Stieltjes transform of $M(t)$ is given by

$$M^*(s) = \frac{F^*(s)}{1 - F^*(s)}, \tag{A.5.2}$$

where $F^*(s)$ is the Laplace-Stieltjes transform of $F(t)$.

Example A.5.1 If we assume that $F(t)$ obeys a gamma distribution, i.e., $X_j \sim GAM(\lambda,\ k)$ in general (k is a positive integer), we have the Laplace-Stieltjes transforms of $F(t)$ and $M(t)$, respectively:

$$F^*(s) = \left(\frac{\lambda}{s+\lambda}\right)^k, \tag{A.5.3}$$

$$M^*(s) = \frac{\left(\frac{\lambda}{s+\lambda}\right)^k}{1 - \left(\frac{\lambda}{s+\lambda}\right)^k} = \frac{1}{\left(1+\frac{s}{\lambda}\right)^k - 1}. \tag{A.5.4}$$

Let $\varepsilon_r = e^{\frac{2\pi r i}{k}}$ be all the distinct roots of an equation $s^k = 1$, where $r = 0, 1, 2, \cdots, k-1$. By applying the partial fraction expansion, we have

$$M^*(s) = \frac{1}{k}\sum_{r=0}^{k-1} \frac{\lambda\varepsilon_r}{s + \lambda(1-\varepsilon_r)}. \tag{A.5.5}$$

By inversion, we have

$$M(t) = \frac{\lambda t}{k} + \frac{1}{k}\sum_{r=1}^{k-1} \frac{\varepsilon_r}{1-\varepsilon_r}\left[1 - e^{-\lambda t(1-\varepsilon_r)}\right], \tag{A.5.6}$$

which has been given in Example 4.2.3.

Example A.5.2 If we assume that

$$F(t) = 1 - \frac{1}{2}\left(e^{-\lambda t} + e^{-2\lambda t}\right), \tag{A.5.7}$$

we have the Laplace-Stieltjes transform of $F(t)$:

$$F^*(s) = 1 - \frac{1}{2}\left(\frac{s}{s+\lambda} + \frac{s}{s+2\lambda}\right). \tag{A.5.8}$$

The Laplace-Stieltjes transform of $M(t)$ is given by

$$M^*(s) = \frac{F^*(s)}{1 - F^*(s)}$$
$$= \frac{3\lambda s + 4\lambda^2}{s(2s + 3\lambda)}$$
$$= \frac{4\lambda}{3s} + \frac{1}{9} \cdot \frac{3\lambda/2}{s + 3\lambda/2} \tag{A.5.9}$$

by applying the partial fraction expansion. By inversion (i.e., applying the formulas in Table A.2), we have

$$M(t) = \frac{4\lambda t}{3} + \frac{1}{9}\left(1 - e^{-\frac{3}{2}\lambda t}\right). \tag{A.5.10}$$

Example A.5.3 If we assume that $F(x)$ obeys a negative binomial distribution of order 2 (refer to Section 2.3), we have the following Laplace-Stieltjes transform:

$$F^*(s) = \left(\frac{pe^{-s}}{1 - qe^{-s}}\right)^2. \tag{A.5.11}$$

The Laplace-Stieltjes transform $M^*(s)$ of the renewal function $M(x)$ is given by

$$M^*(s) = \frac{F^*(s)}{1 - F^*(s)}$$
$$= \frac{pe^{-s}}{2}\left[\frac{1}{1 - e^{-s}} - \frac{1}{1 - (1 - 2p)e^{-s}}\right]$$
$$= \sum_{n=1}^{\infty} \frac{p}{2}\left[1 - (1 - 2p)^n\right] e^{-s(n+1)}, \tag{A.5.12}$$

which implies the (discrete) renewal density

$$m(x) = \frac{p}{2}\left[1 - (1 - 2p)^{x-1}\right] \quad (x = 2, 3, \cdots). \tag{A.5.13}$$

That is, the (discrete) renewal function is given by

$$M(x) = \sum_{j=2}^{\infty} m(j)$$
$$= \frac{px}{2} - \frac{1}{4} + \frac{(1 - 2p)^x}{4} \quad (x = 2, 3, \cdots) \tag{A.5.14}$$

(see Problems 4.4 and 4.5).

Appendix B

Answers to Selected Problems

Problems 1

1.2

(i) AB^cC^c, (ii) $A \cup B \cup C$, (iii) $AB \cup AC \cup BC$,

(iv) ABC, (v)$AB^cC^c \cup A^cBC^c \cup A^cB^cC$,

(vi) $AB^c \cup AB^cC \cup A^cBC = (AB \cup AC \cup BC) - ABC$,

(vii) $A^cB^cC^c$, (viii) $(ABC)^c$.

1.5

(i) $P\{A \mid B\} = 1/3$, (ii) $P\{B \mid A\} = 1/4$, (iii) $P\{A - B\} = 1/4$,

(iv) $P\{B - A\} = 1/6$.

1.9

(i) $(3x^2 - 2y)^3 = 27x^6 - 54x^4y + 36x^2y^2 - 8y^3$,

(ii) $(4x + 3y^2)^3 = 64x^3 + 144x^2y^2 + 108xy^4 + 27y^6$.

1.10

$$\binom{50}{1}\binom{49}{3} = 921,200 \text{ ways.}$$

1.12

(ii) 27 ways.

1.13

(ii) 20 ways.

1.14

n^r (distinguisable), $\binom{n+s-1}{s}$ (indistinguisable).

1.15

$$\binom{13}{5}\binom{13}{3}\binom{13}{3}\binom{13}{2}\Big/\binom{52}{13}.$$

Problems 2

2.7

n	Theoretical values	Data
1	18.27	14
2	13.81	12
3	10.44	12
4	7.89	8
5	5.96	7
6	4.51	4
7	3.41	2
8	2.57	4
9	1.95	4
10	1.47	3
11	1.11	2

2.14

(i) $E[X] = \dfrac{1}{\lambda_1 + \lambda_{12}}, \qquad Var(X) = \dfrac{1}{(\lambda_1 + \lambda_{12})^2},$

(ii) $Cov(X,Y) = \dfrac{\lambda_{12}}{(\lambda_1 + \lambda_{12})(\lambda_2 + \lambda_{12})(\lambda_1 + \lambda_2 + \lambda_{12})},$

$$\rho(X,Y) = \frac{\lambda_{12}}{\lambda_1 + \lambda_2 + \lambda_{12}}.$$

Problems 3

3.3

(i) $t = 0 \quad (9:00\ a.m.), \qquad t = 9 \quad (6:00\ p.m.),$

$$E[N(9)] = \lambda t = 90 \text{ persons},$$
$$Var(N(9)) = \lambda t = 90 \text{ persons}^2.$$

(ii) $P\{N(0.5) = 0\} = e^{-5} = 0.00674.$

3.4

(i) $P\{N(30) = 0\} = e^{-5}, \qquad P\{N(30) = 1\} = 5e^{-5},$

$$P\{N(30) = 2\} = \frac{25e^{-5}}{2}.$$

(ii) Upper limit $90 + 3\sqrt{90} = 118.3$ persons.

Lower limit $90 - 3\sqrt{90} = 71.7$ persons.

3.10

$m(10) = 440$ persons, $\qquad Variance = 440$ persons^2.

3.11

$F\{X_1 \le t\} = 1 - e^{-(\alpha t)^\beta}$ (Weibull distribution).

Problems 4

4.3

$$Var(N(t)) = 2M * M(t) + M(t) - [M(t)]^2$$
$$= \frac{\lambda t}{4}\left(1 - e^{-2\lambda t}\right) + \frac{1}{8}\left[1 - (1 + 2\lambda t)e^{-2\lambda t}\right] - \frac{1}{16}\left(1 - e^{-2\lambda t}\right)^2.$$

4.4

(i) $F^*(s) = pe^{-s}/(1 - qe^{-s}),$

(ii) $M^*(s) = pe^{-s}/(1 - e^{-s}), \qquad M(m) = pm \quad (m = 1, 2, \cdots).$

4.5

$$M(m) = \frac{pm}{2} - \frac{1}{4} + \frac{(1-2p)^m}{4} \qquad (m = 2, 3, \cdots).$$

Problems 5

5.1

(i) $\pi(2) = [0.68 \quad 0.32]$, $\pi(4) = [0.6672 \quad 0.3328]$, $\pi(8) = [0.6667 \quad 0.3333]$.

(ii) $\pi(2) = [0.66 \quad 0.34]$, $\pi(4) = [0.6664 \quad 0.3336]$, $\pi(8) = [0.6667 \quad 0.3333]$.

(iii) $\pi(n) = [2/3 \quad 1/3] \quad (n = 2, 4, 8)$.

5.2

$$\text{(iii)} \quad \mathbf{P}^n = \begin{bmatrix} 1 & 0 & 0 \\ 1-\left(\frac{2}{3}\right)^n & \left(\frac{2}{3}\right)^n & 0 \\ a_n & 1-\left(\frac{1}{2}\right)^n - a_n & \left(\frac{1}{2}\right)^n \end{bmatrix},$$

where

$$a_n = 1 + 2\left(\frac{1}{2}\right)^n - 3\left(\frac{2}{3}\right)^n \qquad (n = 1, 2, 3, \cdots).$$

5.4

$$\text{(ii)} \qquad \mathbf{P} = \begin{matrix} 0 \\ 1 \\ 2 \end{matrix} \begin{bmatrix} 0 & 1 & 0 \\ 1/4 & 1/2 & 1/4 \\ 0 & 1 & 0 \end{bmatrix}.$$

(iii) One class $\{0, 1, 2\}$, recurrent and aperiodic.

5.5

$$\text{(ii)} \qquad p_{i,i-1} = \frac{i^2}{N^2}, \qquad p_{ii} = 2\frac{i(N-i)}{N^2}, \qquad p_{i,i+1} = \frac{(N-i)^2}{N}$$

$$(i = 0, 1, 2, \cdots, N), \qquad p_{i,i-1} + p_{ii} + p_{i,i+1} = 1.$$

(iii) One class $\{0, 1, 2, \cdots, N\}$, recurrent and aperiodic.

5.8

$\mathbf{P}_1$: $\{0, 1, 2\}$ recurrent.

$\mathbf{P}_2$: $\{1\}$ recurrent (absorbing), $\{0, 2\}$ transient.

$\mathbf{P}_3$: $\{0, 1\}$ recurrent, $\{2\}$ recurrent (absorbing), $\{3\}$ transient.

5.9

$\mathbf{P}_1$: $\{0,1,2\}$ recurrent, periodic with period 3.

$\mathbf{P}_2$: $\{0,1,2\}$ recurrent, aperiodic.

$\mathbf{P}_3$: $\{0,1,2,3,4\}$ recurrent, aperiodic.

5.11

(iii) $\mathbf{P}_1$: $\pi_0 = \pi_1 = 1/2$.

$\mathbf{P}_2$: $\pi_0 = \pi_1 = 1/2$.

$\mathbf{P}_3$: $\pi_0 = \pi_1 = 1/2$.

(iv) $\lim_{n\to\infty} \mathbf{P}_i^n = \begin{bmatrix} 1/2 & 1/2 \\ 1/2 & 1/2 \end{bmatrix} \qquad (i = 1,2).$

(v) $\lim_{n\to\infty} \frac{1}{n}\sum_{i=1}^{n} \mathbf{P}_3^i = \begin{bmatrix} 1/2 & 1/2 \\ 1/2 & 1/2 \end{bmatrix}.$

5.12

$$\lim_{n\to\infty} \mathbf{P}^n = \begin{bmatrix} 1/3 & 2/3 & 0 & 0 & 0 \\ 1/3 & 2/3 & 0 & 0 & 0 \\ 0 & 0 & 1 & 0 & 0 \\ 0 & 0 & 0 & 3/5 & 2/5 \\ 0 & 0 & 0 & 3/5 & 2/5 \end{bmatrix}.$$

5.13

$2' \to 0, \quad 4' \to 1, \quad 0' \to 4, \quad 1' \to 5, \quad 3' \to 2, \quad 5' \to 3.$

$$\mathbf{P} = \begin{array}{c} 0 \\ 1 \\ 2 \\ 3 \\ 4 \\ 5 \end{array} \begin{bmatrix} 1/3 & 2/3 & 0 & 0 & 0 & 0 \\ 1/2 & 1/2 & 0 & 0 & 0 & 0 \\ 0 & 0 & 2/3 & 1/3 & 0 & 0 \\ 0 & 0 & 1/2 & 1/2 & 0 & 0 \\ 0 & 1/4 & 0 & 0 & 1/4 & 1/2 \\ 0 & 0 & 3/4 & 0 & 1/4 & 0 \end{bmatrix}.$$

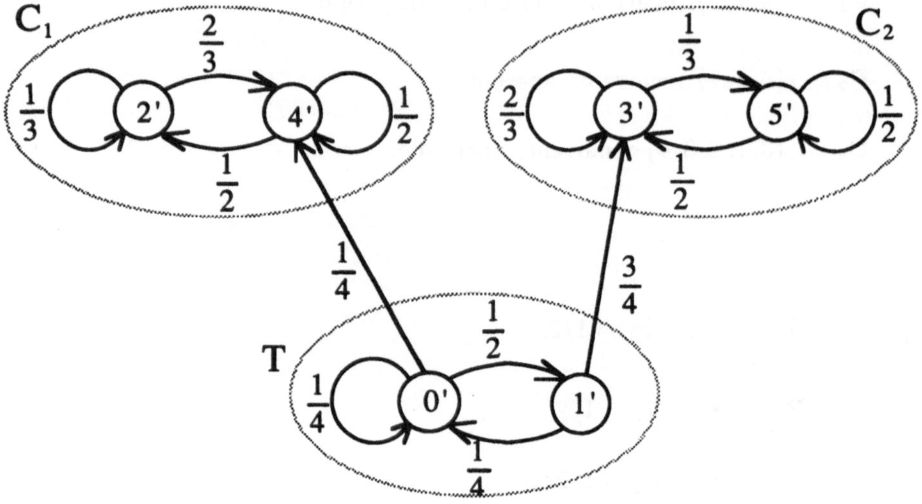

A state transition diagram of Problem 5.13.

$$\lim_{n\to\infty} \mathbf{P}^n = \begin{array}{c} 0 \\ 1 \\ 2 \\ 3 \\ 4 \\ 5 \end{array} \begin{bmatrix} 3/7 & 4/7 & 0 & 0 & 0 & 0 \\ 3/7 & 4/7 & 0 & 0 & 0 & 0 \\ 0 & 0 & 3/5 & 2/5 & 0 & 0 \\ 0 & 0 & 3/5 & 3/5 & 0 & 0 \\ 6/35 & 8/35 & 9/25 & 6/25 & 0 & 0 \\ 3/70 & 4/70 & 27/50 & 18/50 & 0 & 0 \end{bmatrix}.$$

5.15

$$\mathbf{P} = \begin{array}{c} 0 \\ 1 \\ 2 \\ 3 \end{array} \begin{bmatrix} 0 & p & 0 & q \\ q & 0 & p & 0 \\ 0 & q & 0 & p \\ p & 0 & q & 0 \end{bmatrix}.$$

Doubly stochastic (irreducible, recurrent, periodic with period 2).

5.16

$$\lim_{n\to\infty} \mathbf{P}^{2n} = (1 - \frac{p}{q}) \begin{bmatrix} 1 & 0 & \frac{p}{q^2} & 0 & \frac{p^3}{q^4} & 0 & \cdots \\ 0 & \frac{1}{q} & 0 & \frac{p^2}{q^3} & 0 & \frac{p^4}{q^5} & \cdots \\ 1 & 0 & \frac{p}{q^2} & 0 & \frac{p^3}{q^4} & 0 & \cdots \\ 0 & \frac{1}{q} & 0 & \frac{p^2}{q^3} & 0 & \frac{p^4}{q^5} & \cdots \\ \vdots & \vdots & \vdots & \vdots & \vdots & \vdots & \end{bmatrix}.$$

$$\lim_{n\to\infty} \mathbf{P}^{2n+1} = (1-\frac{p}{q})\begin{bmatrix} 0 & \frac{1}{q} & 0 & \frac{p^2}{q^3} & 0 & \frac{p^4}{q^5} & \cdots \\ 1 & 0 & \frac{p}{q^2} & 0 & \frac{p^3}{q^4} & 0 & \cdots \\ 0 & \frac{1}{q} & 0 & \frac{p^2}{q^3} & 0 & \frac{p^4}{q^5} & \cdots \\ 1 & 0 & \frac{p}{q^2} & 0 & \frac{p^3}{q^4} & 0 & \cdots \\ \vdots & \vdots & \vdots & \vdots & \vdots & \vdots & \end{bmatrix}.$$

Problems 6

6.1

$$M(t) = e^{\lambda t}.$$

6.2

$$E[N(t)=n] = \frac{1}{\lambda} + \frac{1}{2\lambda} + \cdots + \frac{1}{n\lambda},$$

$$Var(N(t)=n) = \frac{1}{\lambda^2} + \frac{1}{2^2\lambda^2} + \cdots + \frac{1}{n^2\lambda^2}.$$

6.3

$$E[X(t)=k] = \frac{1}{n\lambda} + \frac{1}{(n-1)\lambda} + \cdots + \frac{1}{(n-k+1)\lambda},$$

$$Var(X(t)=k) = \frac{1}{n^2\lambda^2} + \frac{1}{(n-1)^2\lambda^2} + \cdots + \frac{1}{(n-k+1)^2\lambda^2}.$$

6.4

$$M^{(2)}(t) = ie^{2(\lambda-\mu)t}\left\{i + \frac{\lambda+\mu}{\lambda-\mu}[1-e^{-(\lambda-\mu)t}]\right\} \quad (\lambda>\mu),$$

$$Variance = ie^{2(\lambda-\mu)t}\cdot\frac{\lambda+\mu}{\lambda-\mu}\left[1-e^{-(\lambda-\mu)t}\right].$$

6.5

(iii) $$\alpha(t) = \frac{\mu\left[1-e^{(\lambda-\mu)t}\right]}{\mu-\lambda e^{(\lambda-\mu)t}}, \qquad \beta(t) = \frac{\lambda\left[1-e^{(\lambda-\mu)t}\right]}{\mu-\lambda e^{(\lambda-\mu)t}}.$$

(iv) $$\lim_{t\to\infty} P_{10}(t) = \begin{cases} 1 & (\lambda<\mu) \\ \frac{\mu}{\lambda} & (\lambda>\mu) \\ 0 & (\lambda=\mu). \end{cases}$$

6.8

$$\mathbf{A} = \begin{matrix} 0 \\ 1 \\ 2 \end{matrix} \begin{bmatrix} -(\lambda_1+\lambda_2) & \lambda_1 & \lambda_2 \\ \mu_1 & -\mu_1 & 0 \\ \mu_2 & 0 & -\mu_2 \end{bmatrix}.$$

6.9

(i) $M'(t) = a + (\lambda - \mu)M(t), \qquad M(0) = i.$

(ii) $$M(t) = \begin{cases} ie^{(\lambda-\mu)t} + \frac{a}{\mu-\lambda} & (\lambda \neq \mu) \\ at + i & (\lambda = \mu). \end{cases}$$

(iii) $$\lim_{t\to\infty} M(t) = \begin{cases} \frac{a}{\mu-\lambda} & (\lambda < \mu) \\ \infty & (\lambda \geq \mu). \end{cases}$$

6.10

$$p_j = \frac{a(a+\lambda)(a+2\lambda)\cdots[a+(j-1)\lambda]}{j!\ \mu^j} p_0 \qquad (j = 1, 2, \cdots),$$

$$p_0 = \left[1 + \sum_{j=1}^{\infty} \frac{a(a+\lambda)(a+2\lambda)\cdots[a+(j-1)\lambda]}{j!\ \mu^j}\right]^{-1}.$$

6.11

(i) $$\frac{\partial P(t,s)}{\partial t} + \mu(s-1)\frac{\partial P(t,s)}{\partial s} = \lambda(s-1)P(t,s).$$

(iii) $$P(t,s) = \sum_{j=0}^{\infty} \frac{\left[\frac{\lambda}{\mu}(1-e^{-\mu t})\right]^j}{j!} e^{-\frac{\lambda}{\mu}(1-e^{-\mu t})} \cdot s^j,$$

$$P_{0j}(t) = \frac{\left[\frac{\lambda}{\mu}(1-e^{-\mu t})\right]^j}{j!} e^{-\frac{\lambda}{\mu}(1-e^{-\mu t})} \qquad (j = 0, 1, 2, \cdots).$$

(iv) $$\lim_{t\to\infty} P_{0j}(t) = \frac{(\frac{\lambda}{\mu})^j}{j!} e^{-\frac{\lambda}{\mu}} \qquad \text{(Poisson distribution)}.$$

6.12

$$p_j = \frac{1}{j!}\left(\frac{\lambda}{\mu}\right)^j e^{-\frac{\lambda}{\mu}} \qquad (j = 0, 1, 2, \cdots), \quad \text{Poisson distribution.}$$

6.13

(i) $$\mathbf{P}'(t) = \mathbf{P}(t)\mathbf{A} = \mathbf{P}(t)\begin{bmatrix} -2\lambda & 2\lambda & 0 \\ \mu & -(\lambda+\mu) & \lambda \\ 0 & 2\mu & -2\mu \end{bmatrix}.$$

(ii) $$P_{00}(t) = \left[\frac{\mu + \lambda e^{-(\lambda+\mu)t}}{\lambda+\mu}\right]^2,$$

$$P_{01}(t) = 2\left[\frac{\mu + \lambda e^{-(\lambda+\mu)t}}{\lambda+\mu}\right]\left[\frac{\lambda - \lambda e^{-(\lambda+\mu)t}}{\lambda+\mu}\right],$$

$$P_{02}(t) = \left[\frac{\lambda - \lambda e^{-(\lambda+\mu)t}}{\lambda+\mu}\right]^2.$$

(iii) $$\lim_{t\to\infty} P_{00}(t) = \left(\frac{\mu}{\lambda+\mu}\right)^2,$$

$$\lim_{t\to\infty} P_{01}(t) = 2\left(\frac{\mu}{\lambda+\mu}\right)\left(\frac{\lambda}{\lambda+\mu}\right),$$

$$\lim_{t\to\infty} P_{02}(t) = \left(\frac{\lambda}{\lambda+\mu}\right)^2.$$

6.14

(i) $$\mathbf{P}'(t) = \mathbf{P}(t)\mathbf{A} = \mathbf{P}(t)\begin{bmatrix} -2\lambda & 2\lambda & 0 \\ \mu & -(\lambda+\mu) & \lambda \\ 0 & \mu & -\mu \end{bmatrix}.$$

(ii) $$p_0 = \left(1 + \frac{2\lambda}{\mu} + \frac{2\lambda^2}{\mu^2}\right)^{-1}, \quad p_1 = \frac{2\lambda}{\mu}p_0, \quad p_2 = \frac{2\lambda^2}{\mu^2}p_0.$$

Problems 7

7.1

(i) $$\mathbf{Q}(\infty) = \begin{bmatrix} 0.4 & 0.6 \\ 1 & 0 \end{bmatrix}.$$

(ii) $$\rho_0 = \frac{5\mu}{3\lambda+5\mu}, \qquad \rho_1 = \frac{3\lambda}{3\lambda+5\mu}.$$

7.2

(ii) $$\rho_0 = \frac{\mu}{\lambda+\mu}, \qquad \rho_1 = \frac{\lambda}{\lambda+\mu}.$$

7.3

(ii) $\pi_0 = \pi_1 = \pi_2 = 1/3.$

(iii) $\rho_0 = \left(1 + \frac{\lambda}{\mu_1} + \frac{\lambda}{\mu_2}\right)^{-1}, \quad \rho_1 = \frac{\lambda}{\mu_1}\rho_0, \quad \rho_2 = \frac{\lambda}{\mu_2}\rho_0.$

7.5

(ii) $\mathbf{M}^*(s) = \frac{1}{1 - F^*(s)\,G^*(s)} \begin{bmatrix} G^*(s)F^*(s) & G^*(s) \\ F^*(s) & F^*(s)G^*(s) \end{bmatrix}.$

(iii)
$$G^*_{00}(s) = G^*_{11}(s) = G^*(s)F^*(s),$$
$$G^*_{01}(s) = G^*(s), \qquad G^*_{10}(s) = F^*(s),$$
$$\mathbf{P}^*(t) = \begin{bmatrix} P^*_{00}(s) & P^*_{01}(s) \\ P^*_{10}(s) & P^*_{11}(s) \end{bmatrix}$$
$$= \frac{1}{1 - F^*(s)G^*(s)} \begin{bmatrix} 1 - G^*(s) & G^*(s)[1 - F^*(s)] \\ F^*(s)[1 - G^*(s)] & 1 - F^*(s) \end{bmatrix}.$$

Problems 8

8.2

$$P\{\min(X_1, X_2, \cdots X_n) > t\} = \overline{F}_1(t)\overline{F}_2(t)\cdots\overline{F}_n(t)$$
$$= \exp\left[-\sum_{i=1}^{n} r_i(t)\right].$$

8.3

(i) $\overline{F}_1(t)\overline{F}_2(t)\cdots\overline{F}_n(t) = e^{-\sum_{i=1}^{n}\lambda_i t}.$

(ii) $E[X] = \frac{1}{\sum_{i=1}^{n}\lambda_i}, \quad Var(X) = \frac{1}{\left(\sum_{i=1}^{n}\lambda_i\right)^2}.$

8.6

$$M_H(t) = \frac{\lambda\mu}{\lambda+\mu}t - \frac{\lambda\mu}{(\lambda+\mu)^2}\left[1 - e^{-(\lambda+\mu)t}\right].$$

8.7

$$R_0(x,t) = \frac{\lambda}{\lambda+\mu}e^{-\lambda(t+x)} + \frac{\mu}{\lambda+\mu}e^{-\lambda x},$$

$$R_1(x,t) = \begin{cases} \frac{\mu}{\mu-\lambda}\left[e^{-\lambda(t+x)} - e^{-(\lambda+\mu)t}\right] + \frac{\mu}{\lambda+\mu}\left[e^{-\lambda x} - e^{-\lambda(t+x)}\right] & (\lambda \neq \mu) \\ \lambda t e^{-\lambda(t+x)} + \frac{1}{2}e^{-\lambda x} - \frac{1}{2}e^{-\lambda(t+x)} & (\lambda = \mu). \end{cases}$$

$$\lim_{t\to\infty} R_0(x,t) = \frac{\mu}{\lambda+\mu}e^{-\lambda x},$$

$$\lim_{t\to\infty} R_1(x,t) = \frac{\mu}{\lambda+\mu}e^{-\lambda x}.$$

Problems 9

9.1

(i) $P\{\text{The attendant is busy}\} = \rho = 3/4.$

(ii) $L_q = \dfrac{\rho^2}{1-\rho} = 2\dfrac{1}{4}$ persons.

(iii) $W_q = \dfrac{\rho}{\mu(1-\rho)} = \dfrac{1}{20}$ hour = 3 min.

(iv) $P\left\{U > \dfrac{1}{15}\right\} = \dfrac{3}{4}e^{-1} = 0.276.$

9.2

(i) $P\{\text{The secretary is busy}\} = \rho = 0.8.$

(ii) $L = \dfrac{\rho}{1-\rho} = 4$ jobs.

(iii) $W = L/\lambda = 40$ min.

(iv) $P\{U + V \leq 40\} = 1 - e^{-1} = 0.632.$

9.5

(i) $P\{\text{The dentist is busy}\} = 1 - p_0 = 0.6346.$

(ii) $L = 1.423$ *patients*.

(iii) $W_q = 23.64$ min.

9.8

$$u = \frac{\lambda}{\mu} = 1.0, \quad 2.0 \quad (0.1)$$

u	p_4
1.0	0.0154
1.1	0.0201
1.2	0.0262
1.3	0.0324
1.4	0.0396
1.5	0.0479
1.6	0.0563
1.7	0.0658
1.8	0.0747
1.9	0.0848
2.0	0.0950

$$u = \frac{\lambda}{\mu} \leq 1.5 \implies p_4 = 0.0479 < 0.05 \ (5\%).$$

Appendix C

The Bibliography

[1] A. O. Allen, *Probability, Statistics, and Queueing Theory with Computer Science Applications*, Academic Press, New York, 1978.

[2] L. J. Bain, *Statistical Analysis of Reliability and Life-Testing Models: Theory and Methods*, Marcel Dekker, New York, 1978.

[3] R. E. Barlow and F. Proschan, *Mathematical Theory of Reliability*, Wiley, New York, 1965.

[4] R. E. Barlow and F. Proschan, *Statistical Theory of Reliability and Life Testing: Probability Models*, Holt, Rinehart and Winston, New York, 1975.

[5] U. N. Bhat, *Elements of Applied Stochastic Processes*, Wiley, New York, 1972.

[6] K. L. Chung, *Elementary Probability Theory with Stochastic Processes*, Springer-Verlag, New York, 1974.

[7] E. Çinlar, *Introduction to Stochastic Processes*, Prentice-Hall, Englewood Cliffs, New Jersey, 1975.

[8] D. R. Cox, *Renewal Theory*, Methuen, London, 1962.

[9] W. Feller, *An Introduction to Probability Theory and Its Applications*, Vol. I, 3rd ed., Wiley, New York, 1968.

[10] W. Feller, *An Introduction to Probability Theory and Its Applications*, Vol. II, 2nd ed., Wiley, New York, 1971.

[11] B. V. Gnedenko, Y. K. Belyayev, and A. D. Solovyev, *Mathematical Methods of Reliability Theory*, Academic Press, New York, 1969.

[12] P. G. Hoel, S. C. Port and C. J. Stone, *Introduction to Stochastic Processes*, Houghton Mifflin, New York, 1971.

[13] S. Karlin and H. M. Taylor, *A First Course in Stochastic Processes*, 2nd ed., Academic Press, New York, 1975.

[14] S. Karlin and H. M. Taylor, *A Second Course in Stochastic Processes*, Academic Press, New York, 1981.

[15] J. G. Kemeny and J. L. Snell, *Finite Markov Chains*, Springer-Verlag, New York, 1976.

[16] L. Kleinrock, *Queueing Systems*, Vol. I: *Theory*, Wiley, New York, 1975.

[17] L. Kleinrock, *Queueing Systems*, Vol. II: *Computer Applications*, Wiley, New York, 1976.

[18] J. Kohlas, *Stochastic Methods of Operations Research*, Cambridge University Press, Cambridge, 1982.

[19] R. G. Laha and V. K. Rohatgi, *Probability Theory*, Wiley, New York, 1979.

[20] N. R. Mann, R. E. Schafer, and N. D. Singpurwalla, *Methods for Statistical Analysis of Reliability and Life Data*, Wiley, New York, 1974.

[21] S. Osaki, *Stochastic System Reliability Modeling*, World Scientific, Singapore, 1985.

[22] E. Parzen, *Modern Probability Theory and Its Applications*, Wiley, New York, 1960.

[23] S. M. Ross, *Applied Probability Models with Optimization Applications*, Holden-Day, San Francisco, 1970.

[24] S. M. Ross, *Stochastic Processes*, Wiley, New York, 1983.

[25] H. M. Taylor and S. Karlin, *An Introduction to Stochastic Modeling*, Academic Press, Orlando, 1984.

[26] W. A. Thompson, Jr., *Point Process Models with Applications to Safety and Reliability*, Chapman and Hall, New York, 1988.

[27] K. S. Trivedi, *Probability & Statistics with Queueing, and Computer Science Applications*, Prentice-Hall, Englewood Cliffs, New Jersey, 1982.

Index

Q

R

S